Human Factors 2024

Wageningen, The Netherlands
8-9 October 2024

ISBN: 979-8-3313-2275-5

Printed from e-media with permission by:

Curran Associates, Inc.
57 Morehouse Lane
Red Hook, NY 12571

Some format issues inherent in the e-media version may also appear in this print version.

The Royal Institution of Naval Architects
8-9 Northumberland Street
London, WC2N 5DA
United Kingdom

Phone: 020 7235 4622
Fax: 020 7259 5912

publications@rina.org.uk

Additional copies of this publication are available from:

Curran Associates, Inc.
57 Morehouse Lane
Red Hook, NY 12571 USA
Phone: 845-758-0400
Fax: 845-758-2633
Email: curran@proceedings.com
Web: www.proceedings.com

160 YEARS OF HISTORY, KNOWLEDGE AND EXPERIENCE

THE ROYAL INSTITUTION OF NAVAL ARCHITECTS

Human Factors 2024

8-9 October 2024
Wageningen, The Netherlands

TOGETHER WE WILL SHAPE OUR FUTURE

In Partnership With:

MARIN

THE ROYAL INSTITUTION OF NAVAL ARCHITECTS

Human Factors 2024

8-9 October 2024

Wageningen, The Netherlands

CONTENTS

Human Factors 2024
8th – 9th October 2024
Wageningen, Netherlands

PERILS OF THE HUMAN OPERATOR IN MARITIME AUTONOMOUS SYSTEMS

A Frizell BMT, UK
A Weir, UK

SUMMARY

This technical paper explores some of the Human Factors (HF) challenges surrounding the transition to the deployment of highly autonomous vessels, with an emphasis on the Defence domain. The paper explores the significance of the human operator as a safety influencing factor, and the importance of allocation of function. The issues raised will be of relevance to all maritime vessels that incorporate a high degree of autonomy, but still depend on some degree of onboard crewing or offboard control. This paper considers how we might define the levels of autonomy within a complex system, using a warship as an example, and identifies human-autonomy relationships to be avoided. It further considers the cultural changes required to ensure that the challenges of autonomy can be overcome.

1. UNDERSTANDING THE PROBLEM SPACE

The Highly Autonomous Warship

Over the next 20 years, BMT expects to see the increased adoption of offboard systems, coupled with higher levels of automation and high trust systems on board Naval vessels. However, human interaction will still be required to perform certain operational and decision-making functions due to ethical or operational complexity reasons. To support this, highly autonomous naval vessels will need to enable a small number of personnel to conduct operations and make key decisions while the vessel itself will largely operate autonomously with minimum operator interaction. This paper investigates some of the HF aspects of the development of highly autonomous naval vessels.

Defining Automation and Autonomy

Autonomy and Automation are two terms often used in similar contexts. At a simple level, 'automation' simply describes the use of machines and computers that can operate without needing human control. However, this covers a broad array of capability. The ability to operate free from human control need only to refer the specific process to which that automation is applied, defined by the start and end point of the automated process. The process may be short and simple or long and complex, but it will typically exist as part of a larger task or system with which the process must interface, receiving inputs and providing outputs. An automated process will still require periodic maintenance and may still require a degree of oversight from a human operator, even under normal operating conditions. The automated process will also be subject to external forces and pressures from the environment in which it operates, which may require human intervention to maintain system function.

Autonomy, in general parlance, is the ability to act and make decisions without being controlled. At first glance this appears similar to automation, but the key element here is to 'make decisions'. In the context of an automated process, autonomy therefore denotes that it incorporates sufficient Artificial Intelligence (AI) to be able to manage environmental and contextual variability on its own, responding and adapting without human assistance in order to maintain operation within acceptable performance and operating parameters.

Another key aspect of autonomy is the extent to which it is free from human control, which can be considered in two dimensions: time and scope. An automated process will be time bound, with a defined start and end point outside of which external management is required. Autonomy implies a greater longevity of independence, incorporating a level of machine learning such that the system is able to make decisions about system behaviour in future scenarios based on learned experiences, adapting over time. With regards to scope, an automated process may only manage a subset of a much larger and complex overall task. The author argues that true autonomy implies that we are describing more than a simple process; rather we are describing a system that performs a complete task in its own right. To illustrate this latter point, consider automation in a road vehicle. Basic automation may manage a particular control parameter (such as speed control) but this is meaningless outside of the context of the overall driving task. More complex automation may manage all control parameters, but only under certain conditions. A fully autonomous vehicle would need to manage the driving task as a whole.

For the purposes of this paper the terms automation and autonomy are defined as follows:

- Automation is a general term for where a machine carries out a defined process as programmed, where human intervention is not required under normal operating conditions.

- Autonomy is a subset of automation incorporating a level of Artificial Intelligence (AI) such that the machine is able to adapt to changes in the environment in order to maintain system function and make decisions about system behaviour in future scenarios based on previous experiences; and where the automation performs a complete task in its own right.

Defining the overall task in which automation sits may appear simple, but in practice it is often not straightforward. Any system exists to some extent within a larger system, which itself will be part of a yet larger system. Defining a complete task is therefore somewhat subjective and may require defining a boundary that later shifts. This paper explores some of the reasons for potential ambiguity and the challenges associated with this. It draws upon examples of automation within the automotive domain and transposes lessons learned to the context of a highly autonomous naval warship. It further explores the issues surrounding the definition of the boundaries of system capability and the implications for when systems deemed to be operating without human intervention do, in fact, require human intervention.

Crucially, automation relates only to the specified process under control and makes no assertion that external assistance is not required periodically or at the start and end point of the process. Systems will also operate without human assistance only insofar as they remain within tolerable system boundaries. Autonomous systems may be more resilient to changes in the environment or operating context but will still be governed by limits to their adaptive ability, whether or not these limits are known in advance, and will still be bound by the extent of their remit of authority and capability. Should the system/environment/context stray outside of these conditions then the autonomy may not function as intended (if at all) and external intervention will be required for acceptable performance to be maintained. The degree to which human intervention will be possible in such a scenario is not straightforward and will be explored in more detail in the following sections. The critical aspect is the boundary of capability, whether for automation or autonomy, and the context of the wider system in which it sits.

The Goals of Autonomy

It is important to establish the purpose of autonomy in order to provide a frame of reference for determining the success of its implementation. Warships are, by their nature, potentially subject to hostile action, with the accompanying potential for loss of human life. First and foremost, autonomy (and automation) offers the potential to reduce personnel numbers on board and therefore reduce the exposure of personnel to hazardous situations. In that regard, reduction in head count is a goal of autonomy in its own right. This may also lead to reduced costs, although cost reduction may or may not always be seen as a primary driver.

Beyond that, the goals of autonomy become more specific to the particular system in which they are introduced, although it is perhaps universally true that autonomy and automation will be expected to perform at least as well as, or ideally better, than an equivalent system with human operator(s). In that regard the implementation of autonomy will (or certainly should) be designed to play to the strengths of machines whilst minimising the exposure of the system to human weaknesses.

Allocation of Function

An allocation of function analysis is often used to determine how to allocate jobs, tasks, functions, and responsibilities for the system or capability under analysis. The advantages and disadvantages of the task being performed by a human machine is central to the analysis. There have long been efforts to classify human and machine capabilities according to their strengths and weaknesses. This is sometimes referred to as 'humans are better at, machines are better at' (HABA-MABA). Perhaps the most commonly cited of these is the Fitts List, published in 1951 (Fitts, 1951).

Table 1. Fitts List

HABA	MABA
1. Ability to detect a small amount of visual or acoustic energy	1. Ability to respond quickly to control signals and to apply great force smoothly and precisely
2. Ability to perceive patterns of light or sound	2. Ability to perform repetitive, routine tasks
3. Ability to improvise and use flexible procedures	3. Ability to store information briefly and then to erase it completely
4. Ability to store very large amounts of information for long periods and to recall relevant facts at the appropriate time	4. Ability to reason deductively, including computational ability
5. Ability to reason inductively	5. Ability to handle highly complex operations, i.e. to do many different things at once.
6. Ability to exercise judgment	

It is not the intention of this paper to go into the details of HABA-MABA in the current era, but suffice to say that, despite the developments in AI, the Fitts List is remarkably relevant today considering how long ago it was conceived and the subsequent development in machine capability. As much as the HABA list is still broadly true, it may be tempting to think that as machine learning and AI become ever more capable it is inevitable that machines will eventually supersede humans in all capacities. Whilst this may or may not be true, the context of this paper is the highly autonomous warship, not the entirely autonomous warship, and as such considers the shorter-term future where humans are still a critical element, requiring their strengths to be brought to bear. Human strengths and weakness are of course (at least in the absence of human augmentation) unchanging. For example, humans are universally poor at performing repetitive, mundane tasks over extended periods of time. Equally, humans are not suited to monitoring such processes. It may sometimes seem as though humans are viewed as nothing more than an inconvenience by some designers, a source of potential error to be marginalised and protected against lest they be allowed to cause any harm. One of the perils of automation is often that the operator is excluded from the process, but not from the larger task. As previously discussed, processes do not exist in a vacuum, but rather as part of a wider system and capability. The fact that humans are still involved in this wider system testifies to the dependence we still have on human strengths such as ingenuity, adaptability and problem solving. Humans are currently a critical component of most systems, perhaps all systems when viewed at a sufficiently broad level, and without them those systems would surely fail. We depend on humans to keep systems working, effectively and safely; and must ensure that humans are integrated in such a way as to maximise their ability to do so. Automated and autonomous systems that exist as part of a larger socio-technical system must be designed on the basis of this wider context.

Levels of Autonomy

As mentioned previously, ostensibly automated processes rarely operate entirely without human intervention, be that maintenance, error correction, or by providing the link with the wider task/environment/context at the boundaries of automation. It is perhaps useful to first consider autonomy and automation as regarded within the automotive domain. Vehicles provide a context in which there is a (relatively) simple relationship between a single human operator (the driver) and a machine (the vehicle). Automation in the automotive domain is commonly categorised using the 6-level scale devised by SAE International (SEA, 2016). Level 0 represents the lowest level of automation, the driver is in full control with minor assistance. Level 5 represents full autonomy, with the vehicle in full control and the driver essentially just a passenger. Levels 1-4 involve varying degrees of the sharing of control. Strictly speaking Level 5 may be regarded as full automation rather than autonomy necessarily, insofar that the driving task may be considered to be a single journey only. Full autonomy would imply that all the user's driving needs are met.

- Level 0 – Driver is in control. Automation provides limited warnings and momentary assistance.
- Level 1 – Driver in control. Automation provides single control input assistance.
- Level 2 – Driver in control. Automation provides multiple control input assistance.
- Level 3 – Automation in control but driver monitoring required.
- Level 4 – Automation in control. Full vehicle automation, but only in selected environments.
- Level 5 – Automation in control. Full vehicle automation in all environments.

The key point here is the transition between levels 2 and 3, from the driver being in control to the vehicle being in control. Within the automotive world, level 3 is hugely problematic. If we think back to the discussion of the Fitts List (Table 1), humans as monitors is an extremely poor use of human capability. We are essentially saying that the vehicle is theoretically capable of driving itself but is not sufficiently reliable to be depended upon; or that there are known elements of the operational context that the automation is unable to handle. The driver must therefore be on standby to step in at such time as the automation is asked to exceed the limit of its capabilities. This two fundamental problems:

1. Failure of the automation is likely to happen rapidly and at short notice. The driver is therefore asked to step in without any opportunity to reacquaint themselves with the driving task, nor with the benefit of having been fully engaged in the development of the situation in which they are now asked to take control.
2. Failure of the automation is likely to happen in complex, worst-case situations. The driver is therefore being asked to step in at exactly the time where they will be least-plausibly capable of doing so.

For short, discrete tasks where driver engagement can be maintained (such as conducting a parking manoeuvre) it may be practicable for the driver to act as a monitor of the automation. However, Level 3 is pretty much a no-go for any extended driving activity. Any vehicle ostensibly operating at Level 3 in the real world, during normal driving, is likely doing so without the human failsafe it claims to have. Unfortunately, this is not a view seemingly shared by vehicle manufacturers! Good practice instead requires the jump from Level 2 to Level 4. Level 4 still requires periodic driver control, but without the need for the driver to monitor the automation and step in at short notice. Instead, the vehicle and the driver will conduct a formal handover as the vehicle recognizes it is reaching the boundary of its remit. This is not without its challenges in the automotive world, but it is practicable in principle.

2. DISCUSSION

Transposing to a warship context

The SAE system of classifying levels of autonomy has been introduced here as it is generally well known and understood, but it does require some modification when applying to the context of a warship. The European Defence Agency's Safety and Regulations for European Crewless Maritime Systems (SARUMS) group has produced a similar 6-point scale that does this transposition, but it does so on the basis of classifying an autonomous ship in its entirety. This may be best suited to smaller vessels rather than a larger warship and, for reasons that will be discussed later, it is not necessarily suited for the purposes of this paper. As such the SAE scale is carried forward.

The SAE scale is constructed explicitly around the idea of automation of vehicular control inputs. This may be directly applicable to the role of helmsman but little else, whereas there are a multitude of different roles on a warship when autonomy may be introduced. For the purposes of carrying over to a warship context we might therefore redefine the levels as follows:

- Level 0 – Operator is in control. Automation provides alerts and information only.
- Level 1 – Operator in control but automation manages minor elements of the task.
- Level 2 – Operator in control but automation manages significant elements of the task.
- Level 3 – Automation in control but Operator monitoring required.
- Level 4 – Automation in control. The system can be run in fully autonomous mode, but only in selected environments.
- Level 5 – Automation in control. The system is fully autonomous in all environments.

This is presented graphically in Figure 1.

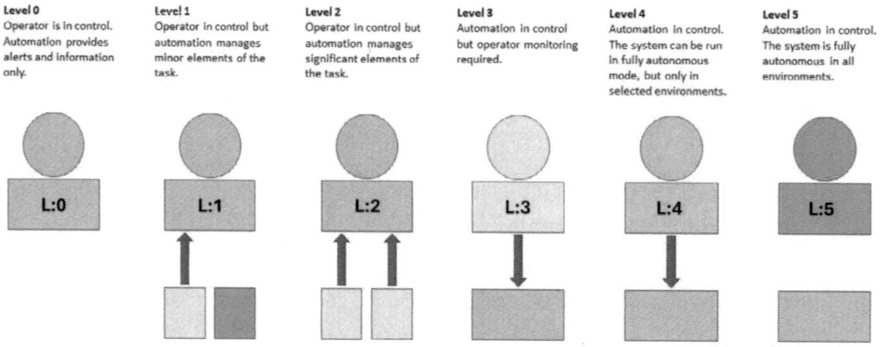

Figure 1. Levels of Automation

Once again, the key transition is from level 2 to level 3, where primary control of the system switches from the operator to the automation. We have already stated that level 3 is generally to be avoided, but in the automotive world level 4 is feasible. When operating at Level 4 the operator is temporarily uninvolved in the process (referred to as being 'out of the loop') whenever the automation has control and must be brought back in the loop before taking back control. The formal handover between vehicle and driver will be directed by the vehicle leaving a defined geographic area or road type for which the automation is certified. This can be planned in advance and will typically present an opportunity for the driver to either get back up to speed with the driving task in advance of taking over, or for the vehicle to pull over if the driver is not ready or not capable of taking over control.

On a warship there are likely to be short notice handovers required in the event of enemy action, and in particular the sustaining of battle damage. Not only may this handover be at short notice, but it likely involves a more complex situation for the operator to understand than a driver might be faced with upon resuming control on a motorway. The operator is therefore asked to step in at the worst possible time, without the benefit of having built up an understanding of the process and its state. This understanding is referred to as situation awareness (SA) and is often cited as having three levels, as defined by Mica Endsley (Endsley, 1995):

1. Level 1 – the perception of environmental elements and events with respect to time or space;
2. Level 2 – the comprehension of their meaning; and
3. Level 3 – the projection of their future status.

SA is not a property of the system, it exists only in the mind of the operator (though is undoubtedly affected by system design). It generally takes time to develop SA, and this is variable dependent on operator experience, state of mind and situation complexity or novelty. Asking an operator to step in at short notice, for a task that is normally automated, means that at least one, and probably several, of the following will be true:

- The scenario is highly complex (given that the automation can't handle it);
- The scenario may be such that previously held assumptions have been violated, and there is no defined correct response;
- The operator may not have been involved in the task prior to stepping in, and lacks even basic baseline SA;
- The operator is de-skilled from lack of practice and experience; and
- This may well be the operator's secondary duty / capability.

The latter two points potentially apply to any situation where there is a level 4 relationship and is something of a paradox as to the benefits of level 4. A system that can act independently for only a short amount of time or in a few select situations limits the potential to reduce operator workload or to allow for operator redeployment. Conversely, a system that can act independently most of the time leads to the operator receiving limited on-the-job experience. This can lead to the operator becoming de-skilled.

In much the same way as in Level 3 for the vehicle, asking an operator to step in under these conditions, especially when that situation is complex and potentially fraught with danger, is perhaps not a realistic prospect. For the purposes of designing systems on a highly autonomous warship then, levels 3 and 4 should be considered as potentially dangerous and to be avoided if possible.

Assessing levels of autonomy in multi-operator systems

The SAE scale (SEA, 2016) is based on a binary relationship of a single human operator in partnership with automation. In multi-operator setups, however, there various relationships that may exist, with different parts of the system operating at different levels of automation. For example, one subsystem may be operating at the equivalent of SAE level 2, whereas another may be at level 4. Take for example, the hypothetical system shown in Figure 2. This would be a fairly typical hierarchical team structure, with a team leader supported by human operators each assigned to the control of individual subtasks. There are multiple operator-automation relationships at play, but none exceeds level 2.

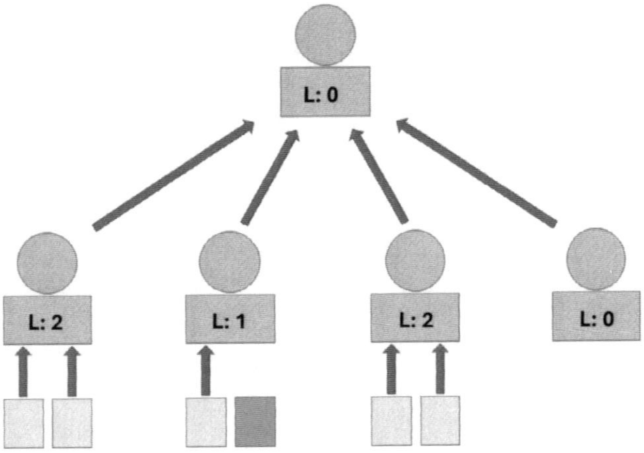

Figure 2. Multi operator- automation relationships at levels 0-2

Consider now the situation in Figure 3. Here, two of the previous roles have been automated, with a single operator monitoring both. Whilst a simplistic example, this represents a possible way that automation may be implemented to reduce crewing levels. The issue in this example is the level 3 relationship that has been created. It is obvious in this case, but may be harder to spot in more complex arrangements. This is particularly so if the designers create what they think is a level 4 or 5 autonomous relationship without due consideration of what the actual task boundary is. As stated previously, full autonomy is not achieved unless the automation can handle the entirety of the task.

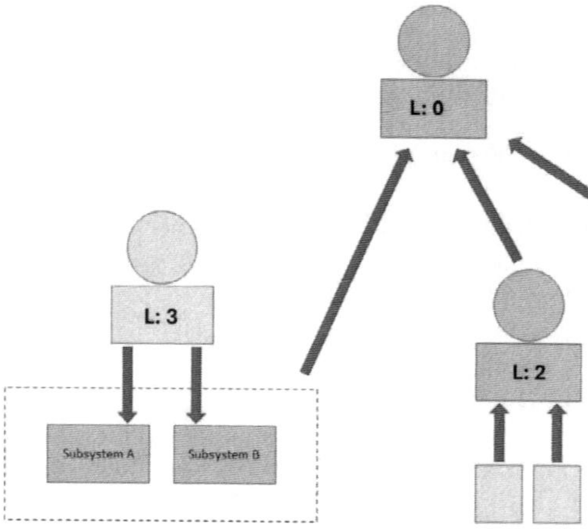

Figure 3. Levels 2 and 3 within the same system

Humans as the third line of defence

The way that humans intervene in a level 3 or level 4 relationship is also potentially altered by automation if the human effectively becomes the third line of defence. Imagine a typical operator intervention scenario where an otherwise automated system goes offline. This operator is likely to be lacking SA to at least some degree and may be under time pressure and/or stress. Not only that, but the operator may first need to spend time understanding where and how they are even needed. As discussed already, this is not an ideal situation. Potentially there could be a separate autonomous recovery system introduced to perform this role in place of the human. If it works then it may solve some of the challenges mentioned, but there is always the possibility that the secondary autonomy is simply a patch that also becomes overwhelmed by the escalating situation. When a human operator eventually has to step in, they do so later in the timeline, with even greater system damage, and lacking the SA they may have developed had they stepped in sooner. Once again, if this is a realistic prospect, the capacity for such an individual to genuinely rescue the situation must be questioned.

Avoiding levels 3 and 4 in complex systems

If levels 3 and 4 are to be avoided, this leaves only two options. Either the system in question remains at level 2 or we make the jump to level 5. On a platform with human operators this jump to level 5 must clearly exist at a lower level than the overall platform. This requires identifying tasks that can meaningfully be automated in their entirety and therefore fully autonomously. This is not to say that the system must be infallible, only that we consider the negatives of any potential failure of the autonomy to be outweighed by the positives of the autonomy. Consider the autonomous car; evidently these will still fail occasionally, and as there is no driver to intervene such failure may be expected to be catastrophic, but the overall benefits are considered to be worth the risk. In the case of the autonomous vehicle a failure may result in a crash. For an autonomous system that exists as part of a wider platform the outcome is not so clear cut. It may be that an autonomous system is sufficiently expendable that the overall ship can continue to function (if potentially in a degraded state). Conceivably it could be the case that a system is so critical that its failure renders the whole ship inoperable, but its failure is determined to be so unlikely, and the benefits of the automation so great, that the cost-benefit trade-off is acceptable. In such an arrangement it is likely that some form of reversionary mode would be planned for, but there must be acceptance that the ability of a reduced crew, with limited experience of this state, to perform in a reversionary mode may be limited at best.

Comparing the performance of humans and autonomy

As stated previously, whether or not it is considered to be the primary reason for introducing autonomy, there will be an expectation that the autonomy performs at least as well as a human undertaking the role. This, however, is not always a straightforward comparison to make. Firstly, because humans and computers have different strengths and weaknesses, they tend to fail in different ways. A good example of this again comes from the world of automobiles and self-driving cars. When a human driver is well trained, engaged and concentrated on the driving task, they can perform to an extremely high standard; capable of responding effectively to highly complex and potentially unfamiliar and unusual situations. Humans tend to fail at the driving task through a lack of attention, either through distraction or fatigue, to the extent that a human driver may even fall asleep at the wheel and therefore have zero attention at all. A self-driving car will never do this. Human failure is often context dependent. A human driver may momentarily fail at a task they would normally be able to do, as a result of their temporary state of mind. Machines, however, tend to be more binary; either they can do something or they can't. Failure of automation will therefore occur when the system exceeds the bounds of its capability, and this can often be dramatic.

There are currently no true level 5 cars, but there are ostensibly level 4 self-driving cars that can at least operate semi-autonomously in defined environments. Statistically, even these relatively early iterations of self-driving cars are safer than human drivers when defined as the rate of crashes per distance travelled. When they do crash though, they tend to make the news. This may partly be explained by them being held to higher levels of scrutiny but also because they often seem to be involved in crashes that any experienced and engaged driver would not be expected to have, often involving a collision with an object that might have been immediately obvious as a hazard to an engaged human driver but simply wasn't detected by the system. Crucially, these objects will typically be unusual in some way. The human brain is incredibly good at recognising patterns and objects and can often tolerate great variability and complexity in the scene. We are typically unaware of what cues we are taking from a scene in order to make our judgments and there may be countless pieces of information our brains are utilising, stitched together from the entirety of our life experiences and understanding of the world. Computers, however, recognise objects by referencing characteristics of that object against previously seen examples of similar objects, using only the (often limited) information channels available to them. Current systems are typically constrained to relatively few recognisable characteristics compared with human recognition. Sometimes, even relatively subtle differences may mean the system fails to recognise that which seems obvious to a human (although this will undoubtedly improve in the future with greater processing capability). In these situations there can be a mismatch between our expectations and reality. We are told that the system is safer than a human driver, and yet it has failed to recognise a hazard that we consider would be obvious to even a novice driver. For many people the overall safety record of the vehicle is then disregarded and only the performance in that specific situation is used to make an appraisal, from which the vehicle does not emerge well. We do not see all the times that the vehicle avoided a crash that a human may not have, only the time it did crash. As long as autonomy fails in ways that are different from typical human failure, there will be a tendency to apportion greater blame in those situations.

Understanding why autonomy can fail

We have so far talked about human and automation 'failure'. This may be interpreted as an implicit apportioning of blame but is intended simply to highlight that the end goal has not been achieved, or has been achieved through unacceptable means or performance parameters. In the case of machines, in particular, this failure to achieve the end goal may not be through any 'fault' of the machine itself. System failures have traditionally been understood through the mechanism of component failure, which propagates through the system into an overall system failure. Modern, complex, computer-based systems may often 'fail' despite working exactly as they were programmed to do. The failure has instead occurred in the design of those systems such that there are scenarios where the programmed behaviour leads to unacceptable outcomes. This may be because of the unexpected emergent properties of complex interactions, or simply because designers made assumptions that proved to be false. Much interesting work has been done in this field by Nancy Leveson of MIT and her development of Systems-Theoretic Accident Model and Processes (STAMP). The point is that complex systems can fail in varied and unexpected ways. If defining a particular system capability as being autonomous we must anticipate that things still can and will go wrong. In such circumstances it will be a human that we depend upon to step in and rectify the situation. The challenge is in planning for this in scenarios that had not been predicted in the first place. The alternative is to accept the consequences when automation failure occurs, but to make this so rare that when it does occur, it is accepted on the basis that the automation has, statistically, outperformed a human operator in the long run. This may require us to constrain the autonomy to applications where we can conceive of a failure being allowed to persist (at least in the short term) without the expectation of human intervention to bring it back online. It will certainly require a change in mentality whereby we judge an autonomous system on its overall performance, not based on specific situations.

Acceptance of human error versus autonomy error

Autonomy may sometimes fail in ways that seem rudimentary to us and may be blamed disproportionately for it as a consequence when viewed purely in the context of that specific situation. But autonomy may also be held to higher standards than human operators, even if failures occur in situations accepted to be challenging even for an expert human operator. When humans make mistakes there is often a tendency to seek out someone to blame, but there is also a degree of understanding and empathy. Provided we feel that the persons involved were genuinely doing their best we at least recognise that people do make mistakes and that challenging circumstances can sometimes prevail despite our best laid plans. Depending on the outcome we may be able to investigate the situation retrospectively to understand what the person was thinking, to understand the decisions they made. This can serve to make the outcome feel more acceptable but also allows us to learn lessons for the future. Sometimes we make changes to the system to convince ourselves (rightly or otherwise) that such an outcome won't happen again. Other times we modify training procedures as mitigation. Indeed, the very presence of established mechanisms for people to be trained and to gain experience and accreditations over time gives us confidence that the system is safe and that failures are isolated cases. This is often not so with autonomy. Machine learning is an immensely powerful tool for allowing computers to generate solutions to otherwise unsolvable problems, but the process can be opaque. We assess the performance of the computer based on the outcome, but we may not necessarily know why it has done what it has. This leads to two issues of trust:

When things go wrong, we may not be able to easily explain why. We cannot ask it what it was thinking or explain it away simply as 'human error' (however inappropriate that may be).

It may be difficult to know how much we should trust the system. There is no time-honoured process for training and easing into service. There is no process for promotion and appraisal; we cannot relate to a time when we were in a similar position, with similar questions and thinking

With regards to point 2, we can go further in saying that we cannot put ourselves inside the mind of a machine. Outside of any specific trade or role, we broadly understand people and how they think, despite our differences. There are so many characteristics of what it is to be human and how we perceive the world. We trust that our colleagues have the ability to adapt and work with us in times of adversity; and that as humans we have at least some ability to display ingenuity in the face of the unexpected, drawing upon a vast life experience outside of the immediate task context. We don't have that with a computer. We can trust it only insofar as we can set it tasks and measure its performance. We cannot know how it will deal with a situation against which it has not been tested and there is no bank of life experience to draw from, only task-specific learning. This is not to say that an autonomous system can never be trusted, only that our frame of reference must be shifted.

3. KEY RECOMMENDATIONS FOR HIGHLY AUTONOMOUS WARSHIPS

The key aspect of the above discussion relates to the importance of correctly defining the system boundary in relation to the task boundary. True autonomy (level 5) can only exist if the system can handle all elements of the task, in all situations. A failure to handle all situations (level 4) means the operator must step in to take control, possibly when under difficult conditions, with limited situation awareness and with potential skill fade. Putting the automation in control but not trusting it (level 3) requires the operator to act as a monitor, which (for anything other than relatively short durations) is something humans are not suited to do.

For a highly autonomous warship, there will be multiple relationships between humans and automation, most likely at varying degrees of human involvement. This all comes together into the complete socio-technical system of the ship and its operating context. The answer is to ensure that the socio-technical system is designed as a whole and not broken up to be designed as individual parts. Human Factors and Engineering must go hand in hand in a coordinated approach. This will be how we ensure that the role of the human within the system is fully accounted for and we are not inadvertently creating situations that the human operators are not equipped to face.

The highly autonomous warship will rely on humans to provide the resilience and adaptability that cannot be fully automated. To do this we must provide those humans with the means to recognise and respond to such scenarios. This means that the system must be designed to provide operators with a means of observing the system state and to recognise when the system is approaching a hazardous state; operators must be able to develop and maintain the situation awareness necessary to understand the implications of that state and what can be done to resolve it; and they must have the skills and capability to enact those interventions. Systems thinking and the associated engineering approaches, adopting multi-disciplinary coordination throughout the CADMID lifecycle, are our best tools to achieving this.

4. CONCLUSIONS

Wherever we still have humans as part of a wider socio-technical system, as is the premise for this paper, there must be acceptance that they are there for a good reason; the system depends upon them to function. Any automation or autonomy that is introduced into that wider system must do so for the benefit of the wider system and the people who are part of it. The automation does not exist in isolation and can be regarded as operating without human control only within the boundaries of the specific function it performs.

Unless a vessel is entirely autonomous it is not useful to classify the entire vessel according to levels of autonomy. Differing levels of autonomy and automation will be achieved for the various systems onboard.

When classifying a system according to its level of autonomy and automation it is necessary to consider the boundary of the system, such that it incorporates the entirety of the meaningful task or function that the system performs. If the system boundary is drawn too narrowly it risks masking a need for operator engagement, situation awareness, experience and skilling.

It is rarely a good idea to reduce a human operator to the role of monitor, based on the previous HABA-MABA discussion. Systems designed on the assumption that a human operator will be there to step in at short notice if required, may be setting the operator up to fail. This will not be the fault of the operator. Systems thinking is required to ensure that system boundaries are drawn appropriately and that potential risks are assessed on a system scale. Human Factors and engineering considerations must be considered together as part of a complete socio-technical system, throughout the Concept, Assessment, Design, In-Service, Disposal (CADMID) lifecycle.

Automation should ideally be introduced only where it supports overall operator control (levels 0-2) or where it is trusted to perform fully autonomously such that no human intervention is assumed, even in the case of failure. It is important to ensure that any relationship between autonomy and humans plays to the strengths of each. The human should not be required to monitor the autonomy, instead the system must be trusted to operate without constant human oversight. If a human will ever be required to take over some or all capability of the autonomy, the human will need to have both the training/experience necessary to perform that role, and must have sufficient time and resources to develop the necessary situation awareness to step in.

The main strength of humans (and likely the reason they continue to be part of the system) is their potential to adapt to unusual circumstances by drawing upon their wider experiences and understanding of the world. To do so requires experience and proficiency; the overall system must facilitate the development of these.

There may need to be a shift in culture as to how we appraise the performance of autonomy. There needs to be an understanding that autonomy may fail in different ways to humans and may do so in ways that appear rudimentary to us. The overall performance of the system must be factored into the assessment. This change in mindset does not only apply within the military, but also to the wider public. Autonomous systems, particularly in the context of the military, attract scrutiny and criticism. Public reaction to high profile failures may be strongly negative and will need to be managed accordingly.

5. ACKNOWLEDGEMENT

BMT would like to thank all those who supported the production of this report and all our customers; without their support none of this would be possible.

The authors would like to especially thank Alistair Weir, Andy Kimber, Russell Bond, Mike Williams, Will Alexander, Aaron McMaster, Chloe Yarrien, Thomas Beard, Eshan Rajabally and Ian Savage for their contributions to the content.

6. REFERENCES

1. ENDSLEY, M.R. Toward a theory of situation awareness in dynamic systems. *Human Factors,* 37(1): 32-64, 1995

2. FITTS, P.M. (Ed.). Human engineering for an effective air-navigation and traffic-control system. *Washington, DC: National Research Council.* (1951)

3. J3016A: Taxonomy and Definitions for Terms Related to Driving Automation Systems for On-Road Motor Vehicles - *SAE International.* (2016, September)

7. AUTHORS' BIOGRAPHY

Alistair Frizell is the Human Factors Capability Lead at BMT. He is Chartered with the Chartered Institute of Ergonomics and Human Factors.

RESPONSIBLE INNOVATION AND REGULATIONS FOR HUMAN SAFETY DURING THE MARITIME TWIN TRANSITION

B S Bondili, **M Lützhöft** and **S N Colclough**, Western Norway University of Applied Sciences, HVL, Norway.
I E Hansen, Norwegian University of Science and Technology, NTNU, Norway.

SUMMARY

Innovation is becoming important in the maritime sector to meet global demands and embrace the green and digital transition. To navigate this twin transition, organisations are focussing on developing and testing new green and digital technologies. The technology that aims to lessen environmental effects and accelerate digitalisation overlooks addressing the crucial question of how safe these innovations are for those using them, notably crew, and seafarers in this sector. The maritime sector depends significantly on regulations to promote safety and innovation, thus the paper will look at the relationship between innovation, regulations, and human safety in this industry during the transition, highlighting the conflicts as well as possibilities for engagement. This study aims to draw attention to the human safety aspects linked to innovation in the maritime industry and address them by integrating the responsible innovation framework and regulations linked to the technology.

1. INTRODUCTION

International shipping carries over 80 percent of the world's merchandise trade by volume, and between 1991 and 2021 the amount of cargo transported by shipping increased from four billion tonnes to eleven billion tonnes [1]. Shipping is a major contributor to transportation, accounting for 3% of greenhouse gas emissions (GHGs) worldwide and using 99% of fossil fuels in the fuel mix [2][3]. The maritime industry is trying to implement greener practices to meet the Net Zero by 2050 set by global agencies, improve competitive positions, and simultaneously reduce the negative effects on the natural environment [4][5][6]. The next industrial revolution is evolving around digital technology, and it is considered that digitalisation can help decarbonise the shipping industry [7]. The two key themes that are currently shaping society are thus digitalisation and sustainability [8]. The Twin Transition is an attempt to promote a sustainable future by utilising digital technologies and greener practices [9]. Voyage optimisations, fleet allocations, port call optimisations, and monitoring the cost and operations are just a few of the many digital support options available to lower greenhouse gas emissions in the maritime sector [10]. It is increasingly acknowledged that digitalisation and decarbonisation intersect in a variety of ways and that managing this 'twin transition' is critical to establishing a sustaining competitive advantage. However, managing the twin transition calls for innovation and change, which presents a significant challenge for organisations. Most maritime organisations, designed for stability and low cost, find it challenging to accept change and differentiation during the twin transition [11].

As the sector strives to meet global demands by embracing the twin transition, ensuring safety is essential as innovations are integrated into maritime operations. Human safety has become a concern since crew, and seafarers are the ones who are likely to be most directly affected by the implementation of innovations and regulatory changes. Safety is a state in which hazards and conditions leading to physical, psychological, or material harm are controlled to preserve the health and well-being of individuals and the community [12]. Safety, as defined in military standard MIL-STD-882E, is "*freedom from those conditions that can cause death, injury, occupational illness, damage to or loss of equipment or property, or damage to the environment*" [13]p.7. "*Safety as the absence of accidents, where an accident is defined as an event involving an unplanned and unacceptable loss*" [14]p.16 [15]. To summarise, safety can be regarded as a state in which there are no unfavourable outcomes, no hazards, and the level of risk is as low as possible. The idea of risk and safety are typically connected in such a way that the larger the risk, the lower the safety, and vice versa. It is insufficient to merely assert that safety is the opposite of risk. However, the relationship between risk and safety has developed further over time, to support effective operations and prevent unforeseen circumstances and uncertainties. The view of safety has in some research traditions moved from Safety-I where the aim is to get the number of unfavourable outcomes as low as feasible to Safety-II, which is the capacity of a system or an organisation to function under a variety of circumstances, hence instead of reducing risk, one increases safety by increasing the number of desired and acceptable outcomes [16].

To increase safety and reduce risk in the maritime industry, innovations and regulations play an important role. In addition to safety and risk management, innovations and amendments to regulations help make this sector sustainable, clean and achieve global targets [17][18]. Some of the measures that drive technological advancements in the maritime industry are energy-efficient propulsion systems; alternative fuels under green technologies; integrated bridge systems, the electronic chart displays as advanced navigation systems; autonomous shipping, the introduction of IoT, smart ship concepts, increased focus on cyber security under digitalisation [19]. Despite the advantages technological innovation brings to this sector and the regulations in place, this industry is still seen as a relatively dangerous place to work [20]. This is perhaps because of a hurried way of introducing technology, haphazard training, confusion during operation, and possible mistakes, ignoring human needs, and capabilities [21][22]. Variability during the development of new technology results in unexpected effects during the implementation and introduction of new technology which may compromising safety [23]. Just as the limitations of technological innovation hinder its benefits, regulatory lapses can limit the growth of innovation in the maritime industry thereby creating unsafe working environments. The lack of regulations and policies in the use of digital technologies creates ambiguities regarding the reporting of potential consequences, such as security breaches, data privacy violations, operational disruptions, and compliance issues in the maritime industry [24]. Similarly, delays in decarbonisation regulation prevent the integration of new technologies and endanger crew safety, especially in the case of the use of alternative fuels [25].

As the industry moves towards decarbonisation and digitalisation, the aforementioned constraints in technology and regulatory shortcomings compromise the safety of individuals who are directly exposed to the innovation. However, human safety does not feature prominently in the discourse on the twin transition, but humans are the ones using the innovations onboard and offshore to bring this transition about. This is why tools that directly emphasise human safety, a socially desirable decarbonisation process, and ethically acceptable digitalisation are necessary during the twin transition. This social desirability and ethical acceptability are central to responsible innovation [26]. "*Responsible Innovation (RI) refers to a transition to a new situation - and an amplification of possibilities - to meet more obligations and honour more duties to fellow human beings, the environment, the planet, and future generations than before*" [27]p.11. The present study aims to address the human safety aspects linked to innovation through the integration of responsible innovation framework and regulations since RI refers to transition, new situations, and meeting more obligations, just like maritime industry in meeting global demands, twin transition, and dealing with innovation.

2. RESPONSIBLE INNOVATION AND HUMAN SAFETY

The creation of new technologies that are in line with social values is at the centre of the philosophy of Responsible Innovation (RI) [28]. RI attempts to prevent unethical, and environmentally harmful outcomes [29]. In the process of innovation, RI encompasses not just accountable organisations but also affected social actors, such as individual members of the society, and the early integration of a variety of stakeholders [30]. RI's goal is not to wait for results and then reflect on them after the fact. Rather, it seeks to either clarify and make explicit the outcomes ingrained in the current technologies or create ways to translate the anticipated outcomes into useful, deployable design parameters [27]. RI directly mentions desirability and acceptability. Then, there can be concerns about how appropriate its framework is for talking about safety. As an explanation for its suitability, the organisations involved in innovation will desire to gain comparative advantages by promoting sustainable practices, and under the practices of ethicality, they consider the safety of the people who are going to get affected by the innovation. The incremental changes that are added to the innovation over the years to gain advantages must not overlook safety. If it is safe to use, it will be sustained, socially desirable, and ethically acceptable. Otherwise, it may fade away along with the organisation's reputation. Therefore, along with acceptability and desirability, safety is also a result of integrating the RI framework into innovation.

In the maritime industry, the RI framework that ensures human safety can be discussed through its four dimensions which are anticipation, inclusion, reflexivity, and responsiveness.

2.1 ANTICIPATION

Unpredictable negative impacts are a common result of new technology, and risk-based damage estimates have frequently fallen short in offering early warnings of potential consequences [31]. Under the anticipation dimension, it is important to acknowledge that the unanticipated effects of novel technologies are not just possible but probable, and there is a need to emphasise uncovering the limits of knowledge rather than proving existing knowledge to be correct. The anticipatory dimension of the decarbonisation process encourages examining the risks to the environment and human health associated with the use of alternative fuels, hybrid propulsion systems, retrofitting to cut emissions, etc. Researchers and organisations are prompted by anticipation to explore contingency, what is known, what is likely, what is feasible, and what is possible by posing "what if" queries [32]. For example, a systematic risk assessment will be anticipated for the introduction of autonomous ships, advanced navigation systems, automation, digital twin generation, and other developments during the digitalisation process to foresee potential risks, failure mechanisms, and their effects on human safety under this dimension.

2.2 INCLUSION

The inclusion dimension calls for involving members of the general public and stakeholder engagement [33]. Inclusion promotes public opinion, involvement, and desirability. The inclusion of seafarers, crew, and organisations involved in the introduction and development of technologies during the transition for feedback and opinions will contribute to making robust systems. Systems can fail, become limited, or have unforeseen consequences when designers and engineers ignore human requirements, capabilities, variability, and a lack of focus on the end user. Under the concept of RI, engineers should move away from looking at their work from a purely technical point of view and ask how their developments may affect humanity's future and whether they are working towards goals of societal improvement and the common good [34]. Enhancing system usability and addressing safety concerns are two benefits of the mentioned stakeholders active participation in the design and implementation phases of new technology. This inclusion feature will promote confidence in using the new technology and mutual trust between organisations attempting to introduce new technology and the seafaring community, who will be implementing this new technology onboard and ashore.

2.3 REFLEXIVITY

Reflexivity is needed to remain aware of one's assumptions, values, role in, and responsibility for society as a researcher or innovator [35]. The reflexivity part of RI aims to better integrate social and environmental issues with financial and economic performance; reflexivity on social and environmental issues should provide guidance rather than slowing the innovation process down or restraining action [36]. Being reflexive, reflecting on one's actions, responsibilities, and presumptions; acknowledging the boundaries of knowledge; and keeping in mind that a specific interpretation of a problem might not be shared by all [28]. As a part of reflexivity, monitoring, and evaluation of new technologies that are introduced during the twin transition should be evaluated for safety performance. The conduct of regular safety audits ensures the reflexivity and these innovations will be strengthened if crew feedback channels are made available for reporting safety issues. Reflexivity impacts the ability of innovators to reflect, drawing links between research practice and regulation, the validity of deliberative exercises, and the formulation of assumptions [28].

2.4 RESPONSIVENESS

The ability to identify and address concerns relating to the three above-mentioned aspects and to alter course in response to them is known as responsiveness [37]. Being responsive indicates that one is prepared to discuss and think about problems that have been voiced, as well as to tentatively accept their validity [36]. Apart from technological and economic considerations, commonly accepted values such as quality of life, and protection of human health & environment serve as anchor points for engagement under the responsiveness dimension in RI [38]. Under this dimension, in the maritime sector, thorough evaluations of new technologies under digitalisation and decarbonisation should take into account the ethical ramifications and effects on human safety. The formation of rapid response teams to address technical malfunctions to reduce hazards and guarantee crew safety. Expanding on the idea of responsiveness, Ayres and Braithwaite (1992) described ways to achieve regulatory compliance, especially in sectors where direct government supervision is difficult [39]. This implies that responsiveness can also be involved in creating systems and processes that encourage adherence to regulations through adaptive and flexible approaches rather than rigid enforcement. As per the Collingridge Dilemma, impacts cannot be easily predicted until the technology is extensively used [40]. This applies to the maritime industry as well.

The presence of Collingridge's dilemma towards the new technology during the twin transition makes it challenging to create regulations for safety issues since these technologies are still in the experimentation stage. However, a lack of regulations should not compromise human safety or the use of technologies; there is a midway that should be approached. Carrying out risk assessments with the help of classification societies, new technologies can be positioned onboard after the nod of the Class. At the same time, innovators can focus on the broad spectrum of regulatory frameworks that exist to support their innovations and meet the safety standards prescribed in the regulations.

3. REGULATIONS IN THE MARITIME INDUSTRY

The regulatory regime in the maritime sector is multi-jurisdictional, multi-layered, and multi-centric among international, regional, national, and local governments, as well as transnational private sector bodies [41]. Although it is multi-layered, IMO has central responsibility and authority to establish global treaties and conventions; private transnational nonstate actors such as the International Association of Classification Societies (IACS), the International Association of Ports (IAP), the International Chamber of Shipping (ICS), and shipping associations such as the World Shipping Council (WSC) play roles in the self-regulation of the sector [41]. The dynamics occurring in the industry in the form of a twin transition call for regulatory initiatives to streamline this transition. This industry lags on regulations such as environmental, security, safety, and economic concerns central to the sector [42]. Probably the reason for the lag in making global regulations is that it is challenging to get a consensus between the 176 member states of this industry and their varied interests in this sector [43]. Additionally, there are economic disparities among countries and their limited roles in investing in digitalisation and decarbonising the industry. Recognising the inadequacy of state response in ratifying conventions and implementing regulations, IMO called for effective implementation in 2014, but the participation of states is still lacking in ratification [44].

By 2050, the International Maritime Organisation (IMO) aims to cut shipping's overall greenhouse gas emissions by at least 50%. Although it offers guidance for strategy in the form of decarbonisation, it is unclear how to set this into practice [45]. To meet the IMO's ambitious goal, the industry focused on technical initiatives such as alternative fuels, clean equipment investments, retrofits, and ship design improvements, as well as other energy efficiency measures, including operational ones. These measures lack regulatory support and have technological constraints. LNG, methyl alcohol, or ethyl alcohol were approved as substitute fuels by the IGF Code, an international code of safety for ships employing gases or other low-flashpoint fuels [46][47]. Under the IGF, no formally agreed-upon international safety rules have been developed for other alternative fuels such as ammonia, hydrogen, biofuels, and batteries. There are safety guidelines available, which come from different independent Classification bodies. The lack of international guidelines on alternative fuels and the varying regulations of Class cause coherence issues in the use of these fuels in the design and operation phases [48]. Regarding technological constraints, these alternative fuels pose a risk to human safety due to their high levels of flammability, toxicity, corrosiveness, explosion potential, and battery shocks. There is an increased danger of mishaps during normal and emergency operations due to implementing innovations without appropriate concerns for users but also a lack of crew training on alternative fuels, modern ship designs, and newly developed materials.

Maritime digitalisation is still in its experimental stage, and regulations regarding autonomous, remote-controlled operations, cyber security, digital twins, the Internet of Things (IoT), augmented and virtual reality, etc., have not been established yet. The introduction of Maritime Autonomous Surface Ships (MASS) with complex technological advancements, which take inputs from predefined algorithms and make decisions based on artificial intelligence (AI), comes under the process of digitalising operations to reduce human errors. The IMO has classified MASS into four degrees of autonomy [49]. Only two of the four categories of autonomous functions are covered by IMO, the other two are left up to the organisations to develop on their own [50]. Cybersecurity became essential after the pandemic due to remote operations and the deployment of IoT devices in the maritime industry. The International Ship and Port Facility Security Code (ISPS), which was established to develop a cooperative mechanism for detecting security threats, and preventing, security incidents affecting ships and port facilities, is not yielding results due to a lack of legally binding mechanisms [51]. These non-legally binding guidelines among member states create inconsistency, non-uniformity, and a threat to cyber security, and ultimately human safety. As threats and technologies evolve, meeting technical criteria is no longer adequate for cyber security. To mandate and obligate the IMO member nations, it is asking for a strong international legal framework, which is currently lacking [51].

The green and digital transition, a twin transition in the maritime industry aimed at decarbonisation and digitalisation through technological innovation, poses challenges in the implementation of these technologies and human safety. Regulations about the safe integration of digital technology and the reduction of carbon emissions are crucial in ensuring the safety of people during this shift. Regulations can ensure that the twin transition is not only technologically and environmentally progressive but also safe for all maritime stakeholders.

4. INTEGRATING RI FRAMEWORK AND REGULATIONS FOR HUMAN SAFETY

The RI framework's reflexivity allows researchers to anticipate potential impacts through the inclusion of stakeholder responses. Industry-specific regulations will guarantee that innovations are applied equitably throughout the industry and that those who violate them are subject to penalties. Regulators should consider changing the rules they established to make them more effective, proportionate, and flexible to meet the demands of emerging technology, particularly during the period of transition. During the early phases of developing guidelines, regulators should seek the participation of stakeholders in the regulatory process. As a participatory approach, stakeholders will provide their valuable feedback based on the responses they received from the people who are working in the sector. For example, in recent times in the maritime industry, IMO's guidelines on maritime cyber risk management called for anticipation, and responsiveness from its stakeholders to make the industry safe and secure against vulnerable threats [52]. The dimensions, anticipation, and responsiveness that are mentioned in the cyber risk management guidelines are the two dimensions of RI, which gives a clear indication that the incorporation of the RI framework will bring safety into the sector.

The concerns related to human safety are addressed through the use of the RI dimensions such as anticipatory risk assessment related to the introduction of new technologies and implementing them onboard, the inclusion of human-centred designs following seafarers' responses, and a reflexive approach towards innovation. During the twin transition, when the integration of the RI framework with the regulations addresses human safety, it can lead to an ethically responsible, socially desirable, and sustainable maritime industry. Technology evaluation is just one example of how RI magnifies long-standing scientific and practical traditions by incorporating elements of anticipation, reflexivity, inclusivity, and responsiveness. This characteristic of technology assessment ensures that the technology will be evaluated in advance rather than after the fact. In the context of RI, the importance of responsibility is viewed as distributed throughout the process as a collective, involving researchers, innovators, funders, policymakers, and other stakeholders such as universities, business/finance, government, or civil society [38][53].

In the maritime industry, stakeholder involvement, covering seafarers, crew, and their safety through the introduction of new technologies onboard, will be improved by RI since responsibility for safety is recognised along the process. As a responsibility by design that is aiming not only technologically advanced but also socially beneficial and ethically sound like RI, this will facilitate stakeholder engagement and interdisciplinary collaborations [54]. For instance, during the twin transition in the maritime sector, the deployment of alternative fuels, autonomous vehicles, and the IoT both ashore and onboard, all of which enhance human-machine interaction. During this interaction, democratising the technology assessment through the inclusion of stakeholders feedback on human-centric designs, emotions on safety-related issues, and finding out regulatory lapses will yield the identification of pros and cons related to the introduction of innovation. As a result, it helps bring technology into existence by making it safe and acceptable to society. The design, development, and use of technology can affect a variety of people, and they may respond to this in different ways. For example, the introduction of digitalisation in the maritime industry facilitates higher connectivity, faster transfer of data from ship to shore, lesser human intervention, and reduced physical storage of data logs and manual entries. But at the same time, it carries a notion that digitalisation will take away the employability of humans. Addressing and managing these conflicts through RI makes them more legitimate since it takes stakeholders views very seriously [55].

5. CONCLUSION

The technological innovation that is taking centre stage in green and digital technologies during the maritime twin transition is linked to operational efficiency, market demands, international guidelines, and human safety to make this industry a sustainable one. In this transitional period, RI contributes a holistic quality to innovation by requiring responsibility at every stage of the process in terms of social desirability, ethical acceptability, and sustainability. Stakeholder involvement, anticipation of future consequences, responsiveness in addressing social and ethical challenges, and reflection on fundamental values all contribute to creating a work environment that fosters productivity without sacrificing security and safety considerations. However, some limiting factors in RI can hinder the expected benefits from the integration of the RI framework and regulations. A few of the factors are the response-ability from stakeholders on the innovation process, such as inherent complexity, uncertainty, and unpredictability of technological innovation on the one hand and moral and political factors that dominate the regulations on the other side in the industry [56]. RI has been criticised for assuming that stakeholders will generally collaborate, align their interests, and aim for consensus [57][58]. The gains that are anticipated from RI could be compromised by limited or absence of stakeholder engagement. Another criticism directed at RI is that it focuses much on upstream factors like anticipating and addressing ethical, social, and environmental issues during the early stages of technology development while ignoring downstream elements like the importance of monitoring, evaluating, and managing the broader impacts that emerge once the technology is deployed [59].

Integrating the RI framework and regulations during the twin transitions in the maritime industry contributes to safety, increased stakeholder engagement, operational efficiency, and adaptability to new technologies. This is because the RI dimensions of anticipation, inclusion, responsiveness, and reflexivity emphasise sustainability, social desirability, and ethical acceptability. Amid the twin transition, operational uncertainties, market demands, technological risk, lack of regulatory support, and seafarers capability to comprehend and apply new technologies onboard present serious obstacles. Nonetheless, the obstacles to placing these technologies into practice will be reduced by proactive stakeholder involvement, risk assessment, training, and government support in the form of new technology incentives.

6. REFERENCES

1. UNCTAD., 'Review of Maritime Transport 2023', *https://unctad.org/publication/review-maritime-transport-2023*, 2023.

2. IMO., 'Third IMO Greenhouse Gas Study 2014', *https://wwwcdn.imo.org/localresources/en/OurWork/Environment/Documents/Third%20Greenhouse%20Gas%20Study/GHG3%20Executive%20Summary%20and%20Report.pdf*, 2014.

3. WEF., (2023). 'Shipping Industry Net Zero Tracker', https://www.weforum.org/publications/net-zero-industry-tracker-2023/in-full/shipping-industry-net-zero-tracker/, 2023.

4. JEGOU, I. S., LAFFINEUR, L., SPIEGELENBERG, F., & LEITÃO, A. M., 'Net-zero by 2050: Achieving Shipping Decarbonization through Industry Momentum and the New Ambition at IMO', *https://unctad.org/news/transport-newsletter-article-no-108-net-zero-by-2050*, 2023.

5. LUN, Y. H. V., LAI, K.-H., WONG, C. W. Y., & CHENG, T. C. E., 'Adoption of Green Shipping Practices in Green Shipping Management (pp. 17-29)', *https://doi.org/https://doi.org/10.1007/978-3-319-26482-0_2*, 2015.

6. SHRIVASTAVA, P., 'Environmental Technologies and Competitive Advantage', *Strategic Management Journal, 16 https://doi.org/https://doi.org/10.1002/smj.4250160923*, 1995.

7. AGARWALA, P., CHHABRA, S., & AGARWALA, N., 'Using Digitalisation to Achieve Decarbonisation in the Shipping Industry', *Journal of International Maritime Safety, Environmental Affairs, and Shipping 5(4), 161-174. https://doi.org/https://doi.org/10.1080/25725084.2021.2009420*, 2021.

8. CASTRO, G. D. R., FERNÁNDEZ, M. C. G., & COLSA, Á. U., 'Unleashing the Convergence amid Digitalization and Sustainability towards Pursuing the Sustainable Development Goals (SDGs): A holistic review', *Journal of Cleaner Production, 280. https://doi.org/https://doi.org/10.1016/j.jclepro.2020.122204*, 2021.

9. MUENCH, S., STOERMER, E., JENSEN, K., ASIKAINEN, T., SALVI, M., & SCAPOLO, F., 'Towards a Green & Digital Future: Key Requirements for Successful Twin Transitions in the European Union', *https://ec.europa.eu/jrc*, 2022.

10. PAKKANEN, P., & VETTOR, R., 'Decarbonization Support from Digital Solutions Providers', *Maritime Decarbonisation, 403-416. https://doi.org/https://doi.org/10.1007/978-3-031-39936-7_29*, 2023.

11. STOPFORD, M., 'Coronavirus, Climate Change & Smart Shipping Three Maritime Scenarios 2020-2050', *S. Maritime. https://www.smm-hamburg.com/fileadmin/presse/pm/sm/2020/Coronavirus__Technology___Smart_Shipping_-_Stopford_20_April_2020.pdf*, 2020.

12. MAURICE, P., LAVOIE, M., LAFLAMME, L., SVANSRÖM, L., ROMER, C., & ANDERSON, R., 'Safety and Safety Promotion: Definitions for Operational Developments', *Injury Control and Safety Promotion, 8(4), 237-240. https://doi.org/https://doi.org/10.1076/icsp.8.4.237.3331*, 2001.

13. DoD., 'System Safety', *Department of Defense USA. https://mail.system-safety.org/Documents/MIL-STD-882E.pdf*, 2012.

14. AVEN, T., 'What is Safety Science', *Safety Science, 67, 15-20. https://doi.org/http://dx.doi.org/10.1016/j.ssci.2013.07.026*, 2014.

15. LEVESON, N., 'A New Accident Model for Engineering Safer Systems', *Safety Science, 42. https://doi.org/doi:10.1016/S0925-7535(03)00047-X*, 2003.

16. HOLLNAGEL, E., 'Safety -I and Safety -II The Past and Future of Safety Management', *(Vol. 1st Edition). CRC Press. https://doi.org/https://doi.org/10.1201/9781315607511*, 2014.

17. LING-CHIN, J., SIMPSON, R., CAIRNS, A., WU, D., XIE, Y., SONG, D., KASHKAROV, S., MOLKOV, V., MOUTZOURIS, I., WRIGHT, L., TRICOLI, P., DANSOH, C., PANESAR, A., CHONG, K., LIU, P., ROY, D., WANG, Y., SMALLBONE, A., & ROSKILLY, A. P., 'Research and Innovation Identified to Decarbonise the Maritime Sector', *Green Energy and Sustainability, 4, 1-14. https://doi.org/https://doi.org/10.47248/ges2404010001*, 2024.

18. SOLAS 1974., 'Amendments to the International Code of Safety for Ships using Gases or other Low Flash Point Fuels (IGF Code), (1974 MSC.458 (101))', *https://wwwcdn.imo.org/localresources/en/KnowledgeCentre/IndexofIMOResolutions/MSCResolutions/MSC.458%28101%29.pdf*, 2019.

19. VIDOVIC, T., SIMUNOVIC, J., RADICA, G., & PENGA, Z., 'Systematic Overview of Newly Available Technologies in the Green Maritime Sector', *Energies, 16(2), 641. https://doi.org/10.3390/en16020641*, 2023.

20. GALIĆ, S., LUŠIĆ, Z., & SKOKO, I., 'The Role and Importance of Safety in Maritime Transportation', *6th International Maritime Science Conference. https://trid.trb.org/View/1423393*, 2014.

21. BHARDWAJ, S., BHATTACHARYA, S., TANG, L., & HOWELL, K. E., 'Technology Introduction on Ships: The Tension Between Safety and Economic Rationality', *Safety Science, 115, 329-338. https://doi.org/https://doi.org/10.1016/j.ssci.2019.02.025*, 2019.

22. BIELIĆ, T., 'Influence of Ergonomic Factors on Ship Managing', *Our Sea: Scientific Journal for the Sea and Maritime Affairs, 51, 172-177. https://hrcak.srce.hr/clanak/12801.*, 2004.

23. LÜTZHÖFT, M., & VU, V. D., 'Design for Safety', *Managing Maritime Safety. Taylor & Francis Group. https://www.routledge.com/Managing-Maritime-Safety/Oltedal-Lutzhoft/p/book/9781138559233*, 2018.

24. KOCA, Y. N., & ERDOĞAN, S., 'Are We Ready for Digitalisation? A Study of the Challenges and Barriers to Digitalisation and Technology Use in the Turkish Maritime Sector', *Polish Maritime Research, 31(2), 132-139. https://doi.org/https://doi.org/10.2478/pomr-2024-0029*, 2024.

25. MASODZADEH, P. G., ÖLÇER, A. I., BALLINI, F., & CHRISTODOULOU, A., 'A Review on Barriers to and Solutions for Shipping Decarbonization: What could be the Best Policy Approach for Shipping Decarbonization?', *Marine Pollution Bulletin, 184. https://doi.org/https://doi.org/10.1016/j.marpolbul.2022.114008*, 2022.

26. HOVEN, J. V D., DOORN, N., SWIERSTRA, T., KOOPS, B.-J., & ROMIJN, H., 'Innovative Solutions for Global Issues', *Springer. https://doi.org/DOI 10.1007/978-94-017-8956-1*, 2014.

27. KORMELINK, J. G., 'Responsible Innovation: Ethics, Safety and Technology', *https://doi.org/ https://doi.org/10.5074/t.2019.006*, 2019.

28. STILGOE, J., OWEN, R., & MACNAGHTEN, P., 'Developing a Framework for Responsible Innovation', *Research Policy, 42(9), 1568-1580. https://doi.org/https://doi.org/10.1016/j.respol.2013.05.008*, 2013.

29. HÄUßERMANN, J. J., MAIER, M. J., KIRSCH, T. C., KAISER, S., & SCHRAUDNER, M., 'Social Acceptance of Green Hydrogen in Germany: Building Trust through Responsible Innovation', *Energy, Sustainability and Society, 13(22). https://doi.org/ https://doi.org/10.1186/s13705-023-00394-4*, 2023.

30. OWEN, R., STILGOE, J., MACNAGHTEN, P., GORMAN, M., FISHER, E., & GUSTON, D., R. OWEN, J. BESSANT, & M. HEINTZ (Eds.)., 'A Framework for Responsible Innovation. Responsible Innovation: Managing the Responsible Emergence of Science and Innovation in Society', *https://doi.org/ https://doi.org/10.1002/9781118551424.ch2,* 2013.

31. EEA., 'Late Lessons from Early Warnings: Science, Precaution, Innovation', European *Environmental Agency EEA, Issue. https://www.eea.europa.eu/publications/environmental_issue_report_2001_22*, 2013.

32. RAVETZ, J. R., 'The Science of 'What-if?', *Futures, 29(6), 533-539. https://doi.org/https://doi.org/10.1016/S0016-3287(97)00026-8*, 1997.

33. STIRLING, A., 'Analysis, Participation and Power: Justification and Closure in Participatory Multi-Criteria Analysis', *Land Use Policy, 23(1), 95-107. https://doi.org/https://doi.org/10.1016/j.landusepol.2004.08.010*, 2006.

34. HANKINS, J., 'What Does "Responsible Innovation" Mean? A New Movement Tries to Define Engineers Roles and Responsibilities in the Innovation Process', *IEEE Spectrum. https://spectrum.ieee.org/what-does-responsible-innovation-mean*, 2015.

35. HOOP, E. D., POLS, A., & ROMIJN, H., 'Limits to Responsible Innovation', *Journal of Responsible Innovation, 3(2), 110-134. https://doi.org/http://dx.doi.org/10.1080/23299460.2016.1231396*, 2016.

36. KOOPS, B.-J., OOSTERLAKEN, I., ROMIJN, H., SWIERSTRA, T., & HOVEN, J.V.D., 'Concepts, Approaches, and Applications', *Springer. https://doi.org/DOI 10.1007/978-3-319-17308-5*, 2015.

37. MACNAGHTEN, P., 'The Making of Responsible Innovation', *Cambridge University Press. https://doi.org/DOI: 10.1017/9781108871044*, 2020.

38. VONSCHOMBERG, R., IN M. OWEN, HEINTZ, & J. BESSANT (Eds.)., 'A Vision of Responsible Research and Innovation', https://doi.org/10.1002/9781118551424.ch3. John Wiley, 2013.

39. AYRES, I., & BRAITHWAITE, H., 'Responsive Regulation', *Oxford University Press. https://academic.oup.com/book/53424*, 1992.

40. COLLINGRIDGE, D., 'The Social Control of Technology', *Georgetown University. https://doi.org/http://hdl.handle.net/10822/792071*, 1982.

41. LISTER, J., 'Green Shipping: Governing Sustainable Maritime Transport', *Global Policy, 6(2), 118-129. https://doi.org/https://doi.org/10.1111/1758-5899.12180*, 2014.

42. ROE, M., 'Maritime Governance and Policy-Making', *Springer London. https://doi.org/https://doi.org/10.1007/978-1-4471-4153-2*, 2013.

43. HACKMANN, B., 'Analysis of the Governance Architecture to Regulate GHG Emissions from International Shipping', *International Environmental Agreements: Politics, Law and Economics, 12, 85-103. https://doi.org/https://doi.org/10.1007/s10784-011-9155-9*, 2012.

44. IMO., 'IMO Conventions: Effective Implementation', *https://www.imo.org/en/MediaCentre/PressBriefings/Pages/04-wmd-launch-2014-.aspx*, 2014.

45. SERRA, P., FANCELLO, G., 'Towards the IMO's GHG Goals: A Critical Overview of the Perspectives and Challenges of the Main Options for Decarbonizing International Shipping', *Sustainability, 12(8), 3220. https://doi.org/10.3390/su12083220*, 2020.

46. IMO., 'IMO 2016: IGF Code: International Code of Safety for Ships Using Gases Or Other Low-flashpoint Fuels', *https://www.imo.org/en/OurWork/Safety/Pages/IGF-Code.aspx*, 2016.

47. IMO., 'IMO 2020: MSC.1/Circ.1621: Interim Guidelines for the Safety of Ships Using Methyl/Ethyl Alcohol as fuel', *https://www.imo.org/en/MediaCentre/MeetingSummaries/Pages/MSC-102nd-session.aspx*, 2020.

48. JANG, H., MUJEEB-AHMED, M. P., WANG, H., PARK, C., HWANG, I., JEONG, B., ZHOU, P., & MICKEVICIENE, R., 'Regulatory Gap Analysis for Risk Assessment of Ammonia-Fuelled Ships', *Ocean Engineering, 287(2). https://doi.org/https://doi.org/10.1016/j.oceaneng.2023.115751*, 2023.

49. IMO., 'IMO Takes First Steps to Address Autonomous Ships', http://www.imo.org/en/mediacentre/pressbriefings/pages/08-msc-99-mass-scoping.aspx, 2018.

50. RINGBOM, H., 'Regulating Autonomous Ships - Concepts, Challenges and Precedents', *Ocean Development & International Law, 50(2-3), 141-169. https://doi.org/https://doi.org/10.1080/00908320.2019.1582593*, 2019.

51. KARIM, M. S., 'Maritime Cybersecurity and the IMO Legal Instruments: Sluggish Response to an Escalating Threat?', *Marine Policy, 143. https://doi.org/https://doi.org/10.1016/j.marpol.2022.105138*, 2022.

52. IMO., 'Guidelines on Maritime Cyber Risk Management MSC-FAL.1/Circ.3/Rev.2', https://www.imo.org/en/OurWork/Security/Pages/Cyber-security.aspx, 2022.

53. GRUNWALD, A., 'Responsible Innovation: Bringing Together Technology Assessment, Applied Ethics, and STS Research', *Enterprise and Work Innovation Studies, 7, 9-31. http://run.unl.pt/handle/10362/7944*, 2011.

54. STAHL, B. C., AKINTOYE, S., BITSCH, L., BRINGEDAL, B., EKE, D., FARISCO, M., GRASENICK, K., GUERRERO, M., KNIGHT, W., LEACH, T., NYHOLM, S., OGOH, G., ROSEMANN, A., SALLES, A., TRATTNIG, J., & ULNICANE, I., 'From Responsible Research and Innovation to Responsibility by Design', *Journal of Responsible Innovation, 8(2), 175-198. https://doi.org/https://doi.org/10.1080/23299460.2021.1955613*, 2021.

55. ROESER, S., & STEINERT, S., 'Emotions, Risk, and Responsibility: Emotions, Values, and Responsible Innovation of Risky Technologies', *Taylor & Francis. https://doi.org/https://doi.org/10.4324/9781003276029*, 2023.

56. BLOK, V., & LEMMENS, P., IN B.-J. KOOPS, I. OOSTERLAKEN, H. ROMIJN, T. SWIERSTRA, & J. V. D. HOVEN (Eds.), 'The Emerging Concept of Responsible Innovation. Three Reasons Why It Is Questionable and Calls for a Radical Transformation of the Concept of Innovation. Responsible Innovation 2', *Springer. https://doi.org/DOI 10.1007/978-3-319-17308-5*, 2015.

57. BLOK, V., 'Look Who's Talking: Responsible Innovation, the Paradox of Dialogue and the Voice of the Other in Communication and Negotiation Processes', *Journal of Responsible Innovation, 1(2), 171-190. https://doi.org/https://doi.org/10.1080/23299460.2014.924239*, 2014.

58. OUDHEUSDEN, M. V., 'Where are the Politics in Responsible Innovation? European Governance, Technology Assessments, and Beyond', *Journal of Responsible Innovation, 1(1), 67-86. https://doi.org/https://doi.org/10.1080/23299460.2014.882097*, 2014.

59. GENUS, A., & ISKANDAROVA, M., 'Responsible Innovation: Its Institutionalisation and a Critique', *Technological forecasting and Social Change, 128, 1-9. https://doi.org/https://doi.org/10.1016/j.techfore.2017.09.029*, 2018.

7. AUTHORS BIOGRAPHY

Bhavana Singh Bondili is currently pursuing her Ph.D. in Nautical Operations at Western Norway University of Applied Sciences, HVL, Norway. She is currently working on Innovations, Regulations, and Human Safety during the Maritime Green Shift.

Margareta Lützhöft holds the current position as a Professor at Western Norway University of Applied Sciences, HVL, Norway, and a Master Mariner. Her research interests include human-centered design and the effects of new technology, with a bearing on maritime safety. She has performed fatigue at sea studies and teaching Human Factors to naval architects. She was the project leader of HUMANE, on the human role in autonomous shipping, and a work package leader in the EU project OCEAN, on navigational safety, led by HVL.

Simon Norheim Colclough is an Associate Professor of innovation studies at Western Norway University of Applied Sciences, HVL, Norway. His research interests lie at the intersection of innovation, cooperation, and entrepreneurship. In recent years he has begun to explore how innovation systems evolve across time and space, integrating insights from economic geography.

Irina Emily Hansen, Researcher at Norwegian University of Science and Technology. She specialises in knowledge management within technology innovation. She earned her Ph.D. in Knowledge Management, focusing on innovation projects that bridge industry and academia. Her expertise lies in managing technology innovation projects, particularly in optimizing production processes. She leverages advanced methodologies, including automated production design, Lean principles, and Industry 5.0, to enhance efficiency and innovation in industrial settings.

Human Factors 2024
8th – 9th October 2024
Wageningen, Netherlands

AN ONLINE, REAL-TIME, COLLABORATIVE, IMMERSIVE TOOL FOR EARLY-PHASE, OPERATION-CENTRIC MARITIME DESIGN

E Gernez, **J E Fauske** and **K Nordby,** Oslo School of Architecture and Design, Norway
E Strange, Kongsberg Maritime, Norway

SUMMARY

Iterative and incremental concept development in maritime applications, particularly ship bridges, necessitates early-phase, operation-centric design due to contextual complexities and limited firsthand experience for designers. We propose an innovative method that reconstructs operational scenarios in virtual reality. We describe the tool we have built over the years to support this approach. We demonstrate the tool's use in the context of user interface concept development for ship docking operations. We show how this approach integrates diverse user perspectives in a detailed operational sequence. The tool employs online collaborative interaction design tools to sketch 2D user interface concepts for docking operations, which are then rendered in 3D for expert collaborative evaluation. We argue that our approach enables rapid iteration across various scenarios, offering cost-effectiveness, accessibility, fidelity, and speed essential for user-centered, early-phase, operation-centric design processes.

1. INTRODUCTION

We present the latest improvements of a design tool we have been using over the years to support user-centred, operation-centric processes [1]. The main application has been the design of user interfaces (UI) for ship bridges in complex operations. In this extended abstract we present the components of the tool, and we illustrate the use of the tool in a design case involving a Pure Car and Truck Carrier (PCTC) ship docking in the harbour of Antwerp.

2. THE CASE

The objective of this case was to propose UI concepts for a satellite-based guidance system for docking large PCTCs. The participants were engineers and designers from Kongsberg Maritime, the company developing the guidance system, mariners and technical experts from a ship owning company, and designers from the Ocean Industries Concept Lab (OICL). The OICL designers proposed concepts based upon the input of the two other companies, and the concepts were evaluated by all the participants. The conceptual work focused on designing UIs that could help bridge workers to gain an immediate understanding of the accurate position of the ship, the distance between the ship and the dock, as well as the ship's velocity vectors. The UIs were to be mediated using Augmented Reality (AR), with the ambition to overlay ship position and velocity vectors over the field of vision of the operators, to help the operators keep their eyes on the ship and the dock, instead of having to look at bridge instruments to obtain the same type of information. In terms of UI design, the question was how to integrate 2D UIs mediated in AR within 3D environments (the central bridge console, and the wing bridge console) that already contained several UIs (the instruments located on both central and wing consoles). The docking operations were modelled with two scenarios: (1) ship approach – bringing the ship towards the dock, in between two mooring lines, and (2) docking – final manoeuvres to dock the ship.

3. THE TOOL

Figure 1 below gives an overview of what the tool can do. In essence, the tool enables to build a 3D environment with which several users can interact in real time, and remotely, either through a computer, or a VR headset. The contents of the 3D environment follow the events of an operational scenario. To do this, the scenario is broken down into steps, and the contents of the 3D environment are adapted to match with what happens in each scenario step. The users can select which scenario step they want to use as backdrop for their exploration and interaction with the 3D environment. This interaction is of two types: (1) all users can move around and experience the 3D environment; and (2) some users can modify a few specific parts of the environment. As a result, the tool enables activities such as presenting a concept in a realistic 3D environment, facilitating user testing, collecting expert feedback, and iteratively modifying a concept with user feedback in real time, in the context of an operational scenario.

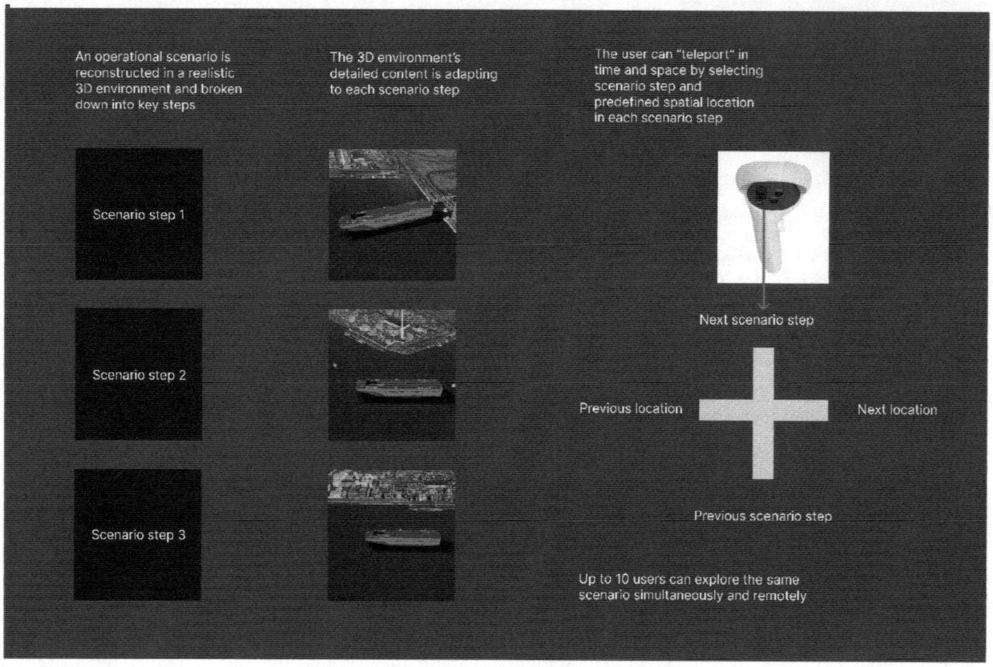

Figure 1: Overview of what the tool can do.

In the ship docking case, we created a scenario where the ship was progressing from the river locks into the docking area in Antwerpen harbour, defining 4 main steps and a total of 12 sub steps (Figure 2). We bought a 3D model of the ship and modelled by hand the details in the deck and bridge required for the scenario. The parts of the 3D environment that could be modified in real-time were the 2D UIs of the ship instruments, and the concepts for UIs to be mediated with AR (Figure 2). In the next sections, we describe how we built the tool to make this type of design approach possible.

Figure 2: scenario steps (top) and example of scenario detail (bottom) in first person view with UI concepts for AR mediated UI and integration of AR-mediated UIs with bridge instruments.

3.1 GAME-ENGINE

The central component of the tool is a game engine, which enables the construction of the 3D environment, and the adaptation of its content to the needs of each step of the operational scenario. We used the game engine Unreal 5 [2], which is available online for free, and has a large community of users and developers. Unreal has powerful visual rendering capabilities that enable to work with highly realistic visuals in real-time, without any latency due to computing time. This is critical to the reconstruction of realistic scenarios and the involvement of external participants. We used two versions of the tool, one made for workshops where participants were able to use standalone VR headsets (i.e not connected to any computer), and one version for high-quality visual production. The two versions are built the same way and use the same assets.

3.2 PLUGINS

Game engines are highly customisable thanks to the use of plugins. Because of its large community of users, Unreal has a vast online library of plugins, with new content created and made available daily by the community.

3.2 (a) Collaborative plugin

Unreal's "collab viewer template" [3] enables up to 10 participants to co-exist in the same 3D environment, and to join remotely from any location. Each participant is represented by an avatar, which enables all the participants to see where they are, where they look, and where their hands point to. This is critical to allow for collaborative discussions. In this case we had participants joining from two different cities in Norway, and one city in Italy. We modified slightly the Unreal native plugin to allow for an optimal use of the VR headset in standalone mode.

3.2 (b) Figma plugin

Digital user interfaces are a critical part of safe and effective operations, and a central part of our contribution to maritime research has been to design user interfaces for maritime workplaces. In this case we accurately recreated all the UIs of the central bridge console and the wing console. Each UI was built in Figma and linked to Unreal using a plugin we developed ourselves. We assigned specific Figma frames to specific scenario steps, and any modification made in Figma to any UI was rendered in the 3D environment immediately. This was also the case for the UI concept designs to be mediated in AR. This enabled a direct loop between design and testing, and we could update and modify the concepts in real-time based on user feedback. This is critical for iterative, user-centred design.

3.2 (c) Ocean and weather plugin

We used the Oceanology plugin [4], and the Ultra Dynamic Sky plugin [5] to render the ocean surface and the sky respectively. Any type of weather can be rendered (calm, storm, etc) at any time of the day (day, night, sunset, sunrise, etc) dynamically. This is critical for creating an immersive experience, but most importantly to test UI concepts mediated in AR in a wide variety of lighting conditions. UIs for AR are notoriously sensitive to backlighting and light contrast [6], so this enabled to quickly evaluate the quality of the different proposed UI concepts.

3.2 (d) Google Maps Cesium plugin

We used the Google Maps Cesium plugin [7], which allows to position 3D assets in any location for which Google Maps has data. In this case we entered the GPS coordinates of the Antwerp harbour, and we instantly got a highly realistic backdrop for the 3D environment. This is particularly useful for the reconstruction of scenarios in coastal areas, to avoid spending time manually modelling critical details such as locks and docks and cranes. We used accurate GPS coordinates for each one of the scenario steps locations.

3.3 DESIGNING WITH THE TOOL

The design case took place over a period of approximately 10 months, through several collaborative sessions, in which different types of design activities took place. In Figure 3 below, we show how the tool is used to support iterative concept design work, and we describe different design activities in the next sections.

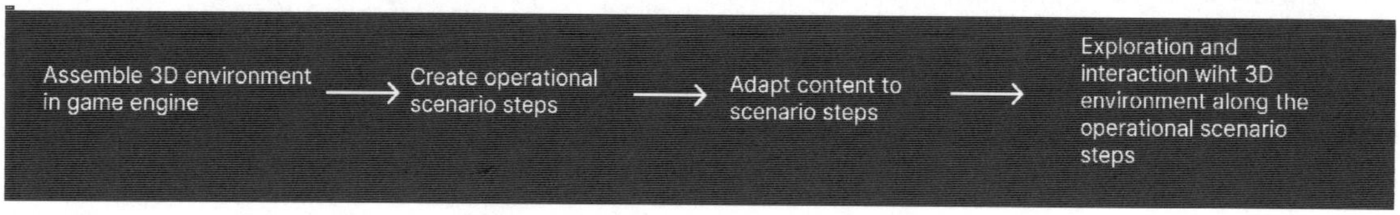

Figure 3: how the tool is set up and used.

3.3 (a) Scenario modelling

We based the scenario on input from the expert participants and on field work – three project participants were able to join a trip on a PCTC docking in Antwerpen harbour. A detailed photographic survey of the bridge was performed and used to update the 3D model of the ship. We performed detailed observations of the navigators during the docking operation: what do they do, where do they stand, where do they look, what instruments do they use, what data from the instruments do they use. A synthesis of this information was put together and, the scenario reconstructed in 3D, and reviewed collaboratively (Figure 4).

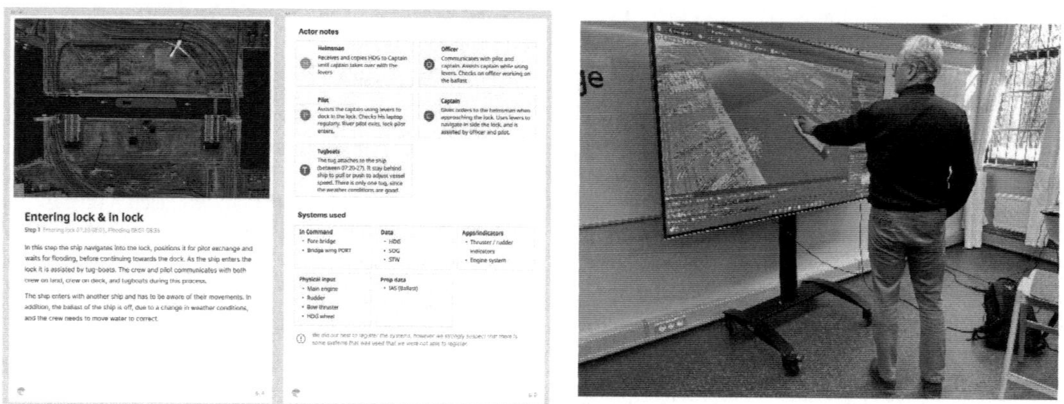

Figure 4: (left) Field observations as input to scenario modelling; (right) Review and completion of scenario

3.4 (b) Design concept presentation and evaluation

The concepts for UIs mediated in AR were first designed in Figma, and then integrated into the 3D environment (Figure 5). A workshop was organised with the expert participants to experience for themselves how the AR-mediated UIs would look and feel like. This enabled to test the concepts in situations where AR design is challenging: lisibility in varying lighting conditions, obstruction of sight and instruments, too much/too little information overlaid over the reality, etc.

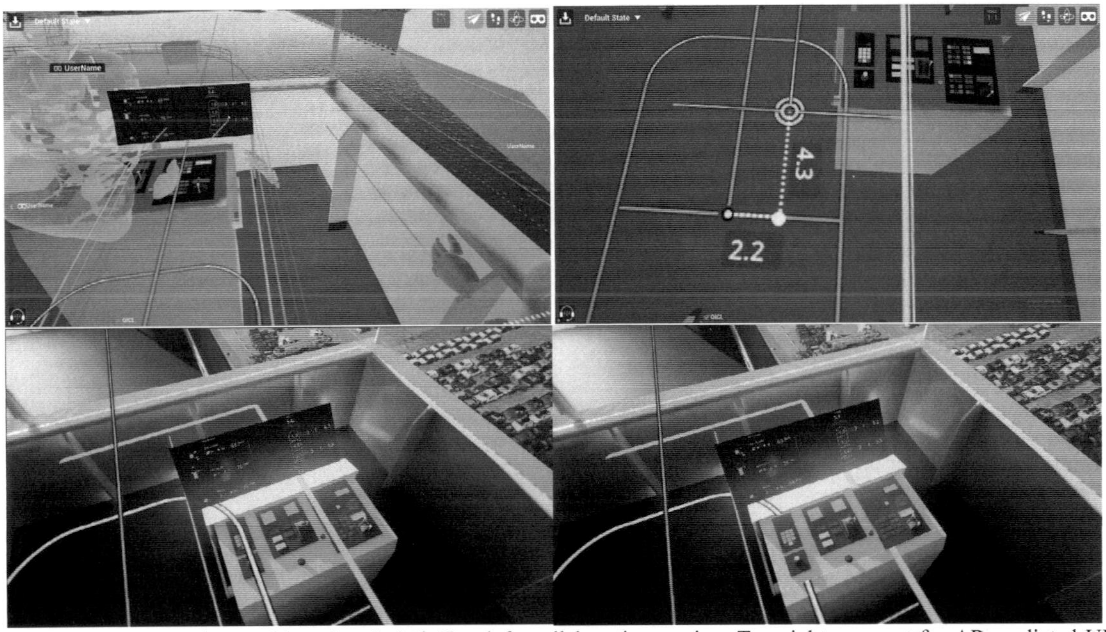

Figure 5: concept review and iterative design. Top left: collaborative session; Top right: concept for AR-mediated UI indicating distance to the dock: Bottom left and right: test, evaluation and correction of an AR-mediated UI with a feature to avoid obstructing the view on critical instruments.

3.4 (c) System testing and fine tuning

Another part of the project investigated how CCTV feeds captured from the dock could be used in the bridge to give navigators a better view of where the ship was in relation to the dock (Figure 6). This required to explore in detail where and how to position the cameras, and how to integrate their feeds with bridge instruments. Virtual cameras were created in the game engines, and different tests were made to find their optimal position. UI concepts for CCTV integration in bridge instruments were developed, some of them also including an overlay of additional information – thus exploring another way to design for augmented reality (with head-mounted displays or on CCTV screens).

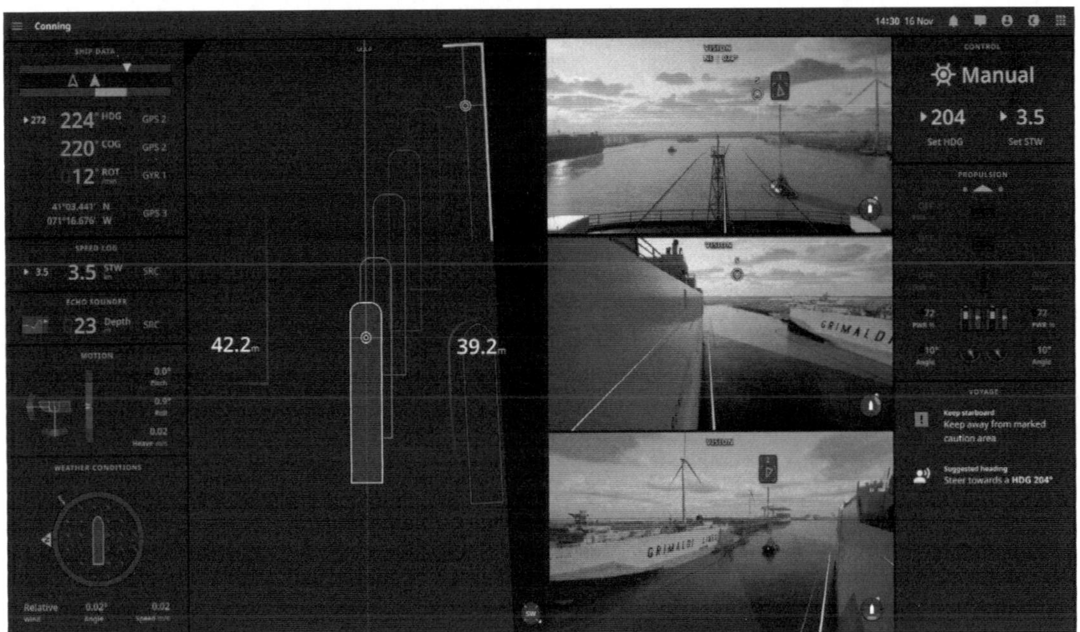

Figure 6: UI concept for integration of CCTV feed in conning UI. Concept presentation made use of virtual cameras to create realistic CCTV feeds

3.4 (d) First person and third person views

In each case, the users participating in VR could chose to experience the scene through a first-person perspective, or they could observe other participants from a distance, from any location in the ship or on the dock. This enabled to test the concepts from many perspectives and cases: a navigator working in the central part of the bridge, teaming with another navigator positioned in the bridge wing; a pilot working with a navigator; a dock-based operator supporting the operations from land and coordinating with the navigators in the bridge. The participants were able to gain a personal experience of each role and perspective and synthesize these experiences in their design process.

4. CONCLUSION

The tool enabled a realistic reconstruction of an operational scenario, in which an iterative, user-centred design process could be facilitated. The tool allowed for great flexibility in the approach, with a high degree of realism, at a low-cost, as every single software used is free and available to all. Some of this work could have taken place in a traditional maritime training simulator, but it would not have allowed for the same flexibility in terms of content creation and adaptation, and for a much higher price tag. Iterative and flexible design are much more challenging to implement in traditional simulators, where content is tailor made to a limited range of scenarios.

5. REFERENCES

1. AYLWARD, K., DAHLMAN, J., NORDBY, K., LUNDH, M., 2021. Using Operational Scenarios in a Virtual Reality Enhanced Design Process. *Education Sciences* 11, 448.

2. Unreal Engine 5 [WWW Document]. URL https://www.unrealengine.com/en-US/unreal-engine-5 (accessed 8.7.24).

3. Unreal engine 4.0 Collab Viewer Template Quick Start [WWW Document]. URL https://docs.unrealengine.com/4.27/en-US/Resources/Templates/CollabViewer/QuickStart/ (accessed 6.9.22).

4. Oceanology in Code Plugins – UE Marketplace [WWW Document]. URL https://www.unrealengine.com/marketplace/en-US/product/oceanology (accessed 8.7.24).

5. Ultra Dynamic Sky in Blueprints – UE Marketplace [WWW Document]. URL https://www.unrealengine.com/marketplace/en-US/product/ultra-dynamic-sky (accessed 8.7.24).

6. FRYDENBERG, S., NORDBY, K., EIKENES, J.O., 2018. Exploring designs of augmented reality systems for ship bridges in arctic waters. Presented at the Human Factors conference, *The Royal Institution of Naval Architects*, London, UK.

7. Cesium for Unreal: Photorealistic 3D Tiles from Google Maps Platform – Cesium [WWW Document]. URL https://cesium.com/learn/unreal/unreal-photorealistic-3d-tiles/ (accessed 8.7.24).

6. AUTHORS' BIOGRAPHY

Etienne Gernez is a senior researcher at the Oslo School of Architecture and Design. His research is focused on the development and use of user-centred design processes for the improvement of workspace design in the maritime industries. He has a PhD in design from the Oslo School of Architecture and Design and previous experience as an engineer in the maritime industry.

Jon Erling Fauske is a designer and researcher in Ocean Industries Concept lab at Oslo School of Architecture and Design. His work is focusing on advanced interaction design for the maritime industries, and he is design lead on the OpenBridge Design System.

Espen Strange is a Senior Designer and Design Researcher at Kongsberg Maritime. His work and research are centered around the design of advanced maritime workplaces, including aspects of industrial design, 3D visualization, human factors, user interface design, and systemic design.

Kjetil Nordby is a professor of interaction design at the Oslo School of Architecture and Design. He manages the Ocean Industries Concept Lab and the OpenBridge Design System. His work primarily focuses on designing advanced maritime workplaces, emphasizing disciplines such as interaction, graphic, and industrial design. For over a decade, he has led numerous research projects in close collaboration with leading maritime industry players.

PROTOTYPE DESIGN AND EVALUATION OF AUGMENTED REALITY FOR COLLABORATIVE SHIP NAVIGATION

F H J Van den Oever, Department of Psychosocial Science, University of Bergen, Norway
B Sætrevik, Department of Psychosocial Science, University of Bergen, Norway
B Orthmann, Floating Lab, Stockholm, Sweden
A van Beek, Troms Fylkestrafikk, Troms County Public Transport Authority, Norway
K Nordby, Ocean Industries Concept Lab, Oslo School of Architecture and Design (AHO), Norway
M Fjeld, Department of Information Science and Media Studies, University of Bergen

SUMMARY

In safety-critical operations like ship navigation, collaboration is key. We developed an augmented reality (AR) prototype for collaborative ship navigation and tested it from shoreside and during voyages. This conference paper reports on the four phases of prototype development, with a preliminary report of the evaluations at the ends of the development phases. Evaluations revealed AR's potential to facilitate collaboration by supporting communication, shared situation awareness, and team decision-making. AR facilitates information gathering, provides an overview, aids mental mapping between 2D charts and the outside view, and increases head-up time. It also supports sharing situation awareness through visual communication. The AR prototype shows promise, particularly in fast operations in medium-busy waterways. There are some challenges related to overreliance, insufficient screen quality, cognitive overload, and sub-optimal usability. Future research should focus on improving accuracy, usability, technology acceptance, and unobtrusiveness to enhance safety and efficiency in safety-critical operations.

1. INTRODUCTION

Collaboration is a critical factor in maritime safety, as evidenced by numerous studies [1, 2, 3, 4]. We adopt the definition of collaboration from Bedwell et al. [5], which describes it as a dynamic process involving two or more social entities that actively and reciprocally engage in joint activities to achieve shared goals. Reviews of maritime accident investigations have revealed various collaboration-related causes [1, 2, 6, 3]. Moreover, maritime accidents and near-misses are frequently underreported [7, 8, 9]. Better collaboration could improve maritime safety.

Concurrently, the maritime industry is evolving. Maritime transport is on the rise [10]. The North Sea is becoming more congested with wind turbines, shipping, solar panels, and aquaculture [11, 12, 13, 14, 15]. Additionally, new technologies, such as (semi-) autonomous and electric-powered vessels, are being introduced [16]. These changes are accompanied by increased risks, particularly as extreme weather events become more frequent due to climate change [17]. Better collaboration can contribute to adapting to climate change. To mitigate climate change, the International Maritime Organization (IMO) has resolved to increase the energy efficiency of ships [18], which can be achieved by improving collaboration [19].

Given the critical role of collaboration in maritime operations for both safety and efficiency, it is imperative to identify new strategies to enhance maritime collaboration in this rapidly evolving industry. In this study, we focus on ship navigation as a representative case of a maritime operation.

2. COLLABORATION

Some of the key elements of collaboration are communication, shared situation awareness (SSA), and team decision-making (TDM). These constructs are frequently examined in the context of maritime operations [3, 20, 21, 22, 23, 24]. Crew members must communicate with each other to build SSA and make decisions [26, 27, 32, 33, 34, 35]. Communication is also necessary for maritime collaboration as a whole [2, 3, 6]. Crew members develop SSA through communicating about SA [32, p. 47]. If a crew has good SSA, they can make good team decisions [30]. Better TDM indicates better collaboration [25, 26, 27, 28, 29]. Within the realm of ship navigation, collaboration encompasses several tasks such as a lookout supporting the navigator with SA, task distribution, navigation hand-over, course planning, and course adjustments [36]. Such collaboration happens more in more complex navigation operations, such as coastal water navigation, dynamic positioning, and bad weather navigation.

Ship bridge crews utilize various information technologies to facilitate collaboration. Enhancements to these communication systems may improve communication and SSA, thereby improving TDM, and thus collaboration. Better collaboration, in turn, contributes to improved safety and efficiency.

2.1 AUGMENTED REALITY FOR SHIP NAVIGATION

This study focused on supporting collaboration during ship navigation using augmented reality (AR). AR supplements the real world with virtual objects that coexist in the same space and time [37]. This technology can be displayed on various hardware such as immersive headsets [38] and fixed displays [39]. It has the potential to improve communication, TSA, and TDM. As such, it can facilitate collaboration in safety-critical operations such as ship navigation [21, 40, 41]. A well-designed AR interface could enhance SA by presenting information where and when the operator needs it [22, 42, 43]. It could also facilitate the mental mapping of 2D screens to the outside world by projecting information onto the outside world [44]. During complex ship navigation, AR could improve SA by reducing head-down time, the time operators spend looking down at screens instead of out of the ship bridge windows [45, 46]. Head-down time may also limit communication between crew members. Studies have demonstrated the positive influence of AR on SA and communication in complex ship navigation tasks [21], ship berthing [47], and sailing [48]. Moreover, AR concepts for icebreaker operations have been developed and tested, suggesting improvements in communication and SSA [38].

Even though research on AR for individual ship navigation is promising [49], research on AR for collaborative ship navigation is still in its early stages and requires further exploration [40, 42, 50]. Furthermore, we argue that research on AR applications in maritime work often lacks grounding in the cognitive and social mechanisms they are intended to support [42]. The impact of AR on factors influencing collaboration is particularly underexplored [21].

AR for ship navigation has primarily been studied through virtual reality (VR) simulations [48, 51], which offer controllable variables, repeatable conditions, and safety [38, 52]. However, for a more ecologically valid understanding of Human-Computer Interaction (HCI) prototypes, it is necessary to study them in real-world settings [53]. This aligns with the need for field research and studying AR for ship navigation at higher Technology Readiness Levels (TRL) to advance research on the topic [42].

2.2 THE CURRENT STUDY

To bring research on AR for ship navigation to the next TRL and higher ecological validity, we iteratively developed a prototype of AR to facilitate collaborative ship navigation. The prototype aims to facilitate collaboration by supporting communication, SSA, and TDM. In the evaluation phases of the iterations, we evaluated the prototype according to these three psychological constructs, as well as design issues. This conference paper presents an intermediate report of this study. It describes the prototype design process, the prototype, and the preliminary methods and findings of the evaluations.

3. THE PROTOTYPE: SJØR

We named the prototype "Sjør", pronounced as "sure." It is a portmanteau of the Norwegian words kjør (to control a vehicle, machine, or the like) and sjø (sea). Developed in Unity using C# [54], it adheres to the OpenXR standard [55]. The interface design aligns with the User Interface Design Framework for Augmented-Reality-Supported Maritime Navigation from the Ocean Industries Concept Lab [41]. Sjør operates on a pair of Microsoft HL2s head-mounted displays (HMDs) and requires a smartphone Wi-Fi hotspot for connectivity to online servers and GPS information. We assess Sjør to be at TRL 7, a system prototype demonstrated in relevant environment [56]. This TRL is higher than most AR prototypes for maritime operations seen in the scientific literature [42]. The code for the prototype is open source and accessible at the provided GitHub repository: https://github.com/thbc/sjoer [57]. The prototype boasts a rich set of features, as listed in Table 1. Figure 1 shows a snapshot of the prototype under usage, labelling the key features. Figure 2 shows a snapshot of the menu. Figure 4 displays a Unified Modelling Language Diagram of Sjør.

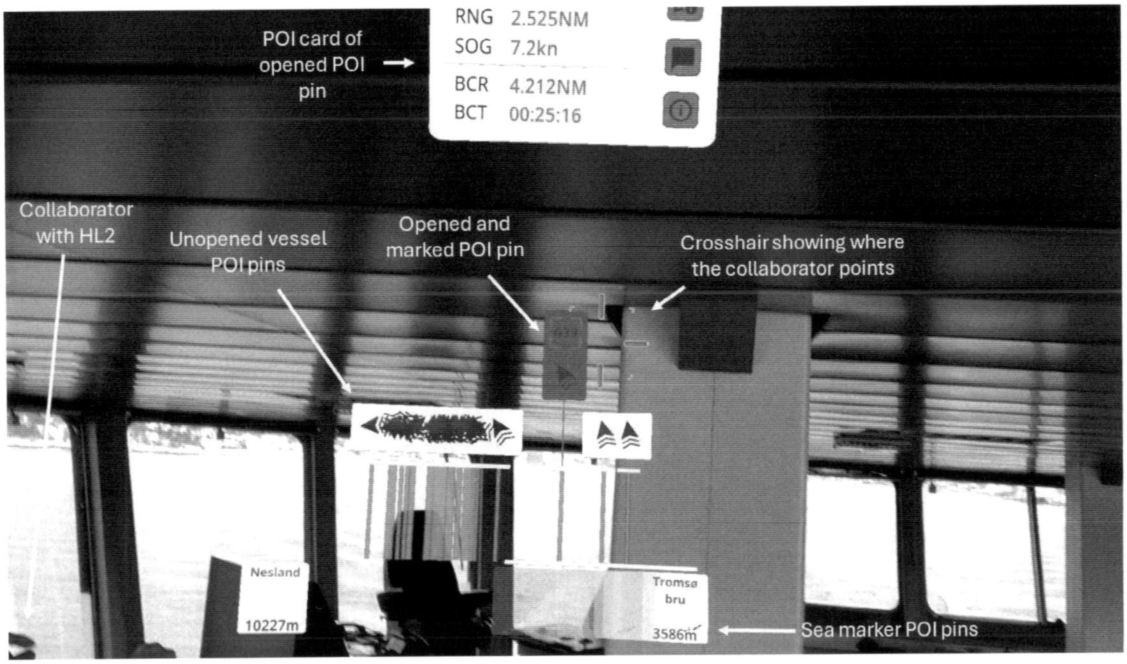

Figure 1: The Sjør prototype in operation. In the centre of the picture are many POI pins of other vessels. One POI pin is opened so the POI card is shown above. Sea marker POIs are shown at the bottom. Just left of the opened POI pin a crosshair shows where the collaborator points with their left hand. At the left bottom of the picture, you can see the left hand of the collaborator.

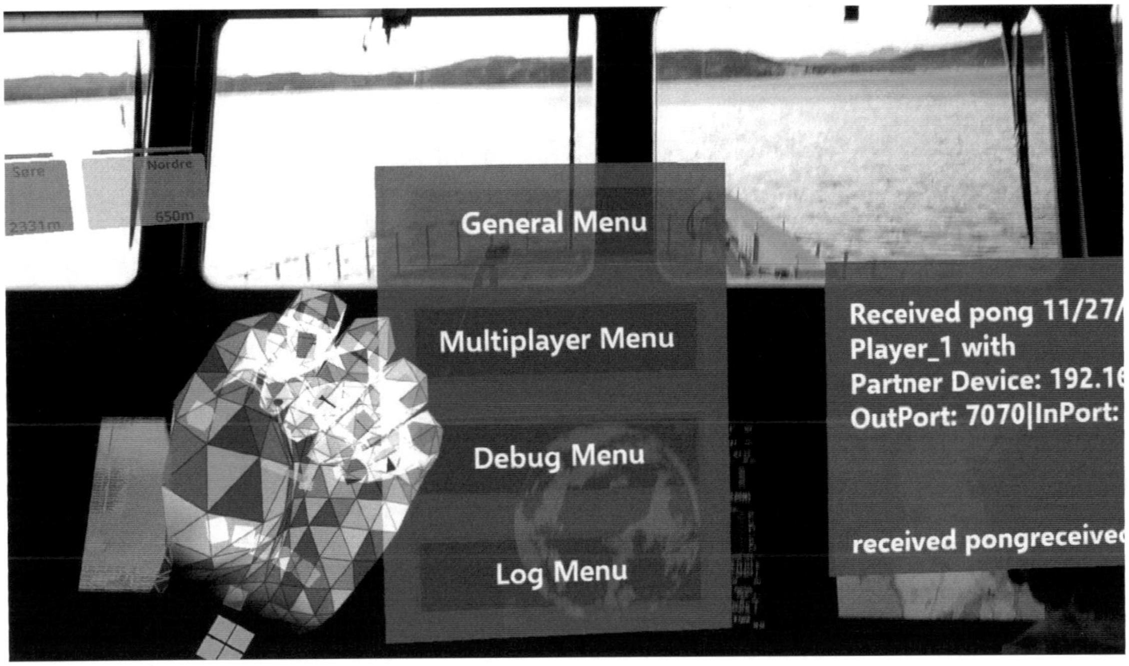

Figure 2: The settings menu. A high-level menu is presented to the user. The menu is opened by holding up one hand. It is operated with the other hand. The information shown to the right indicates the HL2 is connected to the other HL2.

Table 1: Sjør's features, their descriptions, and purposes.

Feature	**Description**	**Purpose**
POI pins for vessels	A visual representation of vessels relative to user GPS position.	To support SA of vessels
POI cards for vessels	A summary of real-time vessel data (based on AIS data).	To support SA of vessels
POI pins for sea markers	A visual representation of navigational aids (NAVAIDS), like lighthouses and buoys relative to user GPS position.	To support SA of NAVAIDS
Eye-tracking	Interaction with pins and cards through gaze-tracking.	To support user experience and information access
Crosshair for pointing with left hand	A visual representation of the collaborator's left-hand pointer.	To support collaboration and SSA
Marking POI pins for vessels with left hand	An operator can colour-mark mark POI pins of vessels for the collaborator. The collaborator can confirm that they received the marking.	To support communication about pins and cards
Graphical User Interface (GUI) with:	A hand-interacting menu to control the GPS connection settings, vessel settings, eye-tracking toggle, brightness, multiplayer, and logging.	To support user control, user experience, and data collection.
GPS app connection settings	Local server IP and port configurations for the device hosting the GPS app that provides the GPS position and orientation to the HL2.	To know the location and orientation of the vessel.
Vessel settings	Settings to adjust the HL2 for the current operation. Range: maximum range that AIS data of vessels and NAVAIDS should be displayed (in m^2). Vessel mode: toggle whether the operation is taking place on a real vessel. Ship name: name of the operator's vessel to skip the vessel's AIS data. Bridge height: height of the vessel's bridge	To support user control and user experience.
Eye-tracking toggle	To enable and disable eye-tracking	To support user control and user experience.
Brightness (night mode)	To iterate through pre-defined brightness settings	To support user experience.
Multiplayer settings	Settings to control the connection between the HL2s. Quick connect: connect to the HL2 last connected with. Connection settings: configure IP and port for the partner device. Enable marking mode: allow operators to mark vessel pins for each other.	To support collaboration and SSA.
Logging	Toggle whether to write parameters to a log.	To collect parameter data of the HL2.

3.1 THE PROTOTYPE DESIGN PROCESS

The study's work was organized as follows. Over the course of 3 years, Sjør was developed adhering to an agile, iterative development process [58, 59], thereby implementing research through design [60]. We roughly followed the interaction design lifecycle model (see Figure 3). On the macro level, we went through the cycle once. On the meso level, we cycled through the model four times. On the micro level, we cycled through the model several times in each of these four phases. In this section, we describe the four meso level cycles.

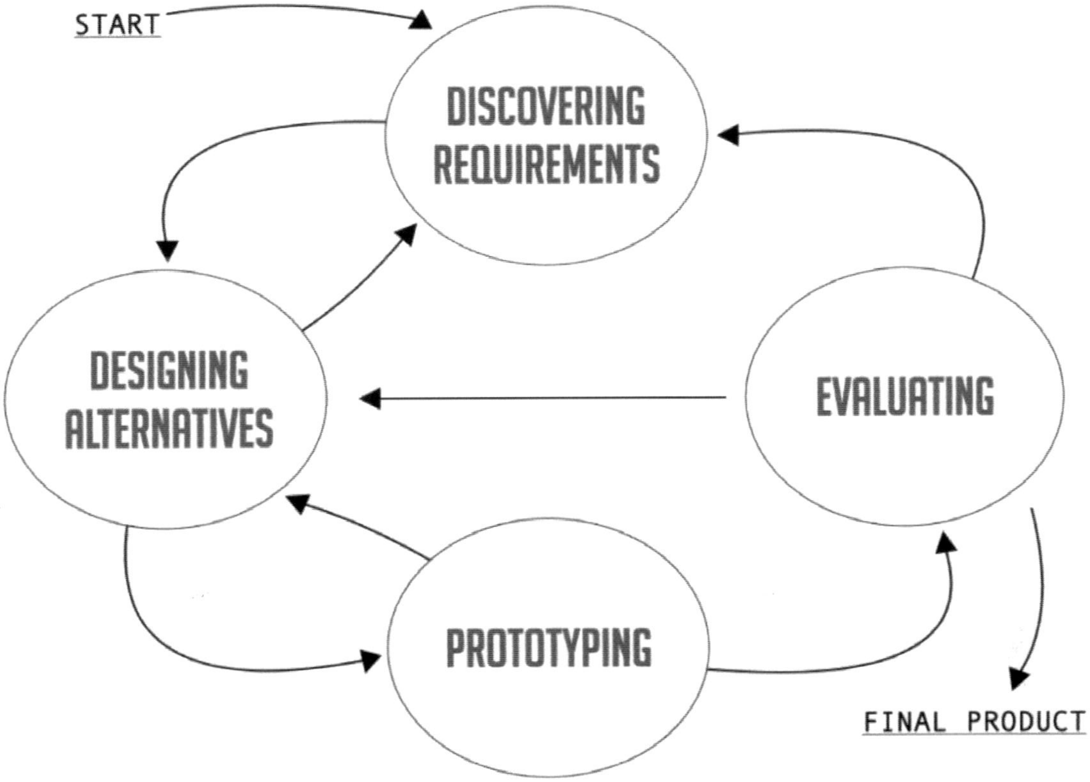

Figure 3: The Interaction Design Lifecycle Model [61].

In the first phase, the prototype's first design was informed by interviews and use case workshops conducted on a platform supply vessel. The platform supply vessel was visited for two consecutive days while it was docked in a port. This phase was the requirements discovery. Requirements discovery in the field allows designers and researchers to get a unique, first-hand understanding of users and their environment. This is essential for a strong foundation for further research and design [62]. Besides that, it can give users insight into the technology under design, which allows them to share better insights with the researchers. Three ship bridge crew members, the captain, and the two first officers, participated. Based on this requirements discovery, two use cases with user stories were worked out of AR applications in two maritime operations: one for coastal water navigation and one for dynamic positioning. Besides that, two additional potential use cases were identified: one for the engine room and cargo room and one for the ship crane [63].

In the second phase, a prototype was designed and developed for the coastal water navigation use case. This was the first version of Sjør. The design and development were mostly performed by the fourth author of this conference paper. The code for the prototype is available on GitHub (https://github.com/imable/sjoer) [63, 64]. This version displays Points of Interest (POI) pins of vessels on a single HoloLens 2 (HL2). These POI pins display the direction the vessel is going. When the user gazes at a POI pin or clicks a POI pin, it opens a POI card with the name, heading, range, speed over ground (SOG), closest point of approach (CPA), and time to closest point of approach (TCPA), as shown in Figure 1. The prototype was tested on a recreational sailing vessel [63].

The third phase involved the discovery of requirements for a multi-user version of the prototype. Requirements were discovered based on two interviews with end users and two workshops with users and designers. The features of pin sharing and custom markers were developed in a Wizard of Oz fashion. This means that these features were testable on one HoloLens pretending to interact with another HoloLens. This prototype is available on GitHub (https://github.com/cheierli/sjoer) [65, 66]. It was tested in a harbour from the shoreside by a ship navigator. An overview and prioritisation of collaborative features for ship navigation AR were made [66].

In the fourth and final phase, the prototype's second version was developed by the third author of this conference paper. The requirements and prototyped features from the third phase were used. This version was designed to support collaborative navigation using two HL2s. It is available on GitHub (https://github.com/thbc/sjoer/) [57]. It has the collaborative features of marking POI pins and a crosshair for pointing. Besides that, it has POI pins of sea markers (such as lighthouses and buoys; Figure 1), and a settings menu (Figure 2). The remainder of this conference paper discusses the second version of the prototype and the field study conducted to evaluate it.

3.2 SYSTEM ARCHITECTURE

The Sjør prototype is structured into distinct modules, organized into six categories: the data pipeline, scene renderer, graphics pipeline, global player & camera object, multiplayer, and global helper function. The system architecture is depicted in the Unified Modelling Language diagram in Figure 4. This section describes the purpose and workings of the modules in the six categories.

The data pipeline module retrieves AIS data from Barentswatch, GPS position, and orientation from the smartphone, and sea marker data from the Norwegian Coastguard [67]. Since the sea markers used in this application are stationary installations, their positions are static. Thus, to reduce network traffic WFS requests are only made if the bounding box area is changed by a defined amount (default 10m²). These data streams are then passed on to the scene renderer as Data Transfer Objects (DTOs). The smartphone logs and sends GPS position and orientation data using Net GPS [68]. The Data Pipeline is set up such that it can also retrieve information from other sources, such as the ship systems via BlueCTRL's Blue Box [69]. If connected to ship systems, Sjør could retrieve more accurate and timely data than provided by the smartphone and online resources. The Scene Renderer processes the DTOs provided by the Data Pipeline into InfoItems, which are Unity Game Objects. This facilitates the user in interacting through eye-tracking, gesture, and voice. Subsequently, the Graphics Pipeline renders these InfoItems as 3D models to display relevant data. The Global Player and Camera Object monitor the current location and orientation of the HL2s in relation to the vessel and relay this information to the Positioner. The Positioner then translates GPS DTOs into Unity coordinates, ensuring that InfoItems are displayed in the correct locations. Global Helper Functions provide utility functions, including unit conversion and mathematical functions for positioning objects. The Multiplayer module, utilizing Open Sound Control, manages all aspects of network connectivity between HL2s within a local Wi-Fi network. This module also establishes connections, enables data exchange, and implements the crosshair and POI pin marking.

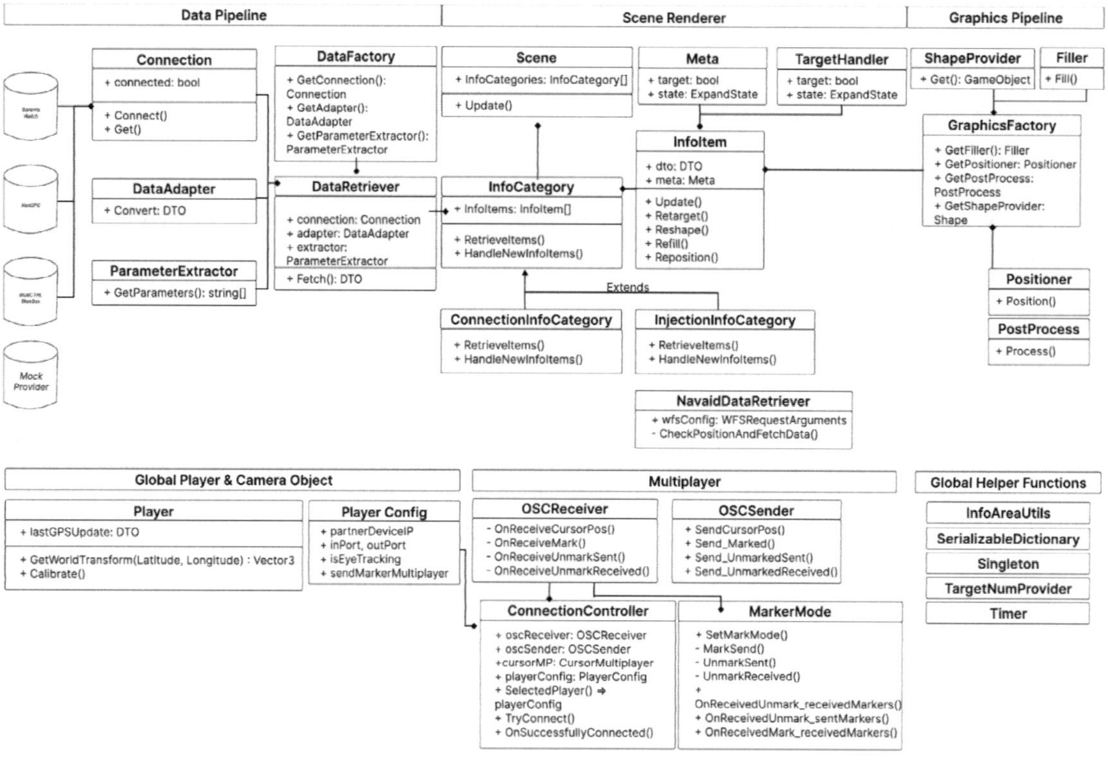

Figure 4: Unified Modelling Diagram of Sjør

3.3 POSITIONING AND ORIENTATING

In order to accurately align virtual POIs pins in a HoloLens 2 (HL2) with the corresponding real-world POIs, each HL2 needs to be aware of its own geographical location and orientation, as well as the locations of the POIs being referred to. This process involves the conversion of real-world Global Positioning System (GPS) position and orientation data into the 3D Cartesian world coordinates utilized by the HL2s. In our prototype, this conversion is performed by a component referred to as the "Positioner", which follows the instructions and algorithm proposed by Van Schaik [70]. This section presents equations that facilitate the transformation of GPS coordinates of POIs, enabling the accurate positioning of POI pins on them.

The user is surrounded by a cylindrical display with a radius of five meters (see Figure 4). The POI pins, POI cards, and crosshairs are displayed on this cylinder, aligned with the POIs they represent. Equation 1 formalizes the horizon plane, which is a cross-section of the cylinder, and defines the cylinder's radius. Equation 2 formalizes the graphics areas on the cylinder. The 3D vector P represents the position of the user and the 3D vector O represents the direction of a POI from the user's perspective. Both vectors are described in the 3D Cartesian coordinate system employed by the HL2, where the distance between two points is measured in real-world meters. For instance, a graphic at the coordinate (0 1 0) is positioned 1 meter above the coordinate (0 0 0). Equation 3 formalizes the rotation matrix that aligns the graphics in the HL2s with geodetic north, ensuring that POI pins and cards on the cylinder align with the POIs they represent. This matrix corrects the offset of the heading of the vessel and the rotation around the Y-axis of the HL2, similar to the approach used by Leite et al. [71]. The position and orientation are continuously updated to accommodate the movement of the vessel, the HL2, and POIs.

$$HorizonPlane = \begin{bmatrix} Px \\ Py - h \\ Pz \end{bmatrix} + dO, where\ h = bridge\ height, d = 5, and\ O = \left(\begin{bmatrix} Ox \\ Oy \\ Oz \end{bmatrix} - \begin{bmatrix} Px \\ Py \\ Pz \end{bmatrix}\right) / ||(\begin{bmatrix} Ox \\ Oy \\ Oz \end{bmatrix} - \begin{bmatrix} Px \\ Py \\ Pz \end{bmatrix})||$$

Equation 1: the horizon plane.

$$GraphicsArea = HorizonPlane + \begin{pmatrix} 0 \\ q \\ 0 \end{pmatrix}, where\ q = the\ graphic\ hight$$

Equation 2: the graphics area.

$$C = -\begin{bmatrix} 0 \\ c + u \\ 0 \end{bmatrix}$$

$$where\ c = vessel\ heading\ at\ calibration - HL2\ roation\ around\ Y\ at\ calibration, and\ u = current\ vessel\ heading - current\ HL2\ rotation\ around\ Y$$

Equation 3: the rotation matrix C.

Figure 5: the cylindrical display

3.3 COMMUNICATION INFRASTRUCTURE

The multiplayer functionality is enabled through User Datagram Protocol based Open Sound Control (OSC), which allows lightweight, low latency network communication out of the box and avoids third-party services. When entering multiplayer mode the system tries to connect to another device within the same local WIFI network with the last used IP address configuration. Choosing between two user identities will set the in- and out-ports to preset values accordingly. Once the connection is initiated, the device sends a "ping"-command out to the entered partner device IP address, while also listening to an incoming "ping"-command from the partner device. When a "ping" is received, the HMD confirms the connection with a "pong"-command. From that moment on, a blue crosshair visualizes in real time where the partner is pointing. If MarkMode is activated, the operator can mark POI pins of vessels for the partner operator by clicking on a pin with the left hand. A network command is then sent to the other device, where the same pin is then rendered in the respective colour, based on the following states: unmarked, received mark, sent mark, and sent and received mark. These states are consistently synchronized between devices.

4. PRELIMINARY REPORT ON THE PROTOTYPE EVALUATIONS

Sjør was evaluated at the ends of the second, third, and fourth phases of the design process. In this paper, we report on the findings of these evaluations. At the end of the second phase, it was tested during recreational sailing and from shoreside. At the end of the third phase, it was tested from the shoreside in a harbor. In the fourth phase, it was tested on three voyages in Norwegian coastal waters.

4.1 METHODS OF THE PROTOTYPE EVALUATIONS

In the second phase, Sjør was tested by one experienced navigator and a designer during two shoreside test sessions and a 1-day sailing trip on a 12-meter sailing boat. In the third phase, Sjør was tested with one experienced navigator and a designer during a shoreside test session. The navigators and designers reflected on design issues of AR for ship navigation as well as requirements for further development.

In the fourth phase, Sjør was evaluated during three multi-day voyages in Norwegian coastal waters. The prototype was tested by ship crews during actual coastal water navigation. Data collection involved interviews, observations, and questionnaires, adhering to a realist epistemology [72]. The study was registered in and approved by the University of Bergen's System for Risk and Compliance. Two voyages were conducted with a 35-meter research vessel near Bergen lasting four and three days respectively. The third voyage, lasting three days, was conducted with a 77.5-meter research vessel from Tromsø to Bergen. A total of 12 participants with varying ship bridge experience tested the prototype, participated in interviews, and completed questionnaires. Data was collected using the HL2s running the prototype, a GoPro camera, a 3D camera, a smartphone, and voice recorders. Questionnaires included the System Usability Scale (SUS) [73], and the Technology Acceptance Model (TAM) [74]. An interview guide structured the interviews during and after testing. The study was conducted by the first author, who ensured the prototype functioned correctly at the start of each voyage. Crew members were invited to participate when they had time in between their normal responsibilities. Participants familiarized themselves with the prototype under the researcher's guidance before testing it and providing feedback. Data collection was flexible to accommodate crew responsibilities.

Qualitative data of the fourth-phase evaluation was thematically analysed [72] using NVIVO [75]. Audio recordings were transcribed automatically with Microsoft Word for the web and checked by the first author and two research assistants. Transcripts and observation notes were coded by the first author and a research assistant following a codebook based on a VR simulator study of AR for ship navigation [51]. A Cohen's Kappa of 0.87 indicated high inter-rater agreement [76]. Disagreements were resolved by the first author. Quantitative data from the SUS and TAM Questionnaire are not included in this conference paper. These will be published in a future publication.

4.2 PRELIMINARY RESULTS OF THE PROTOTYPE EVALUATIONS

Participants suggested that AR could enhance SA by making information more accessible, providing an overview, and reducing head-down time. This could aid in anticipating future situations (level 3 SA) and decision-making. However, participants mentioned that AR could potentially cause attentional capture and limit vision, which could negatively impact SA and decision-making. For AR to be a reliable tool for building and sharing SA, participants stated that it must be highly accurate in both space and time. AR could facilitate collaboration among crew members using AR or a combination of AR and other information systems like ECDIS and radar. Features like the crosshair for pointing and marking pins for collaborators were found useful for sharing SA and TDM. Participants also saw advantages of AR included high usability, operator mobility, and support for mental mapping 2D charts from ECDIS and radar to the 3D outside world.

Disadvantages included insufficient screen quality, physical discomfort, risk for cognitive overload, potential for overreliance, and technology limitations and failures.

The evaluations brought forth various improvement opportunities for AR for ship navigation. Increasing spatial and temporal accuracy is necessary for higher reliability and trustability. Several hardware improvements, like enhancing screen quality are necessary for an acceptable usability and user experience. Participants want more features and more control over these features. Examples of AR features to develop are course lines, collaboration rooms, computer-vision-based detection of objects without AIS, range markers, seabed visualisation, and view sharing. Integrating AR with other information systems will increase its usefulness. More communication support tools will increase its ability to facilitate collaboration. AR developers should consider usability, cognitive load, attentional capture, and overreliance. AR could be particularly useful in medium-busy waterways like fjords, individual navigation, intermittent use for lookouts, high-speed navigation like ferries, and dynamic positioning. Participants also suggested the potential use of AR for shipboard engineering. Quantitative data from the SUA and TAM Questionnaire are not included in this conference paper. These will be published in a future publication.

5. DISCUSSION AND CONCLUSION

Our study suggests that mobile, head-worn AR can facilitate collaborative ship navigation. AR could improve individual SA and communication about SA, thus improving SSA, thereby improving TDM. This aligns with previous research [21, 51, 52]. AR could function with multiple users or with one AR user while others use conventional systems, providing a complementary perspective by facilitating mental mapping of 2D screen information to the outside world. Increased operator mobility, as found by Rowen et al. [22], is another benefit. AR could be particularly useful for individual ship navigation, a focus of previous research [23, 38, 48, 51]. Until AR becomes unobtrusive, it is best suited for intermittent use, serving as a tool for increasing SSA in specific situations. However, potential disadvantages include cognitive overload, visual clutter, attentional capture [51, 77, 78], overreliance [79], and challenges in achieving acceptance and usability. For AR to be truly useful in ship navigation, it must be highly accurate in both space and time, have various features and settings while avoiding visual clutter, attentional capture, and cognitive overload, and be integrated with other ship bridge systems like ECDIS and radar. High usability and technology acceptance can improve safety and efficiency, as found for ship bridge systems in general [80, 81]. These findings may extend to AR for other safety-critical operations such as dynamic positioning of maritime vessels, maintenance and inspection, surgery, and aviation.

Our findings should be interpreted in the light of the limitations of field research. The decision to test a tangible prototype in real-world operations resulted in a trade-off between experimental control and ecological validity. The experimental control in our field test was constrained due to the inability to strictly control testing conditions, repeat tests, evaluate the prototype's performance, and conduct tests with a large sample size, as compared to laboratory studies such as those conducted by Rowen et al. [22]. However, by devising various testing methodologies for the prototype, we were able to navigate some of the unpredictability inherent in field research [82].

We invite both researchers and industry professionals to build on our research. Our prototype is available under a GPL-3.0 open-source license on GitHub: https://github.com/thbc/sjoer/. Further investigation into AR for ship navigation could involve performance testing of higher TRL prototypes, with an emphasis on enhancing spatial and temporal accuracy. Research could also explore the application of AR in other maritime operations such as maritime traffic control and maintenance and inspection [42]. Researchers in other industries, such as medicine and aviation, could consider the potentially generalizable findings from this study. These include the provision of various features focused on SA while avoiding visual clutter, attention capture, and cognitive overload. Maritime industry practitioners could utilize our findings to develop AR for maritime operations, thereby promoting safer, more efficient, and more sustainable work in safety-critical operations such as ship navigation. In particular, AR could reduce the risk of accidents in complex, safety-critical operations that require collaboration [83, 84].

In conclusion, AR can facilitate collaborative ship navigation by supporting communication, SSA, and TDM. Key advantages include increasing information accessibility, supporting mental mapping between 2D charts and the outside view, increasing head-up time, and augmenting communication. AR may be particularly useful in fast operations in medium-busy waterways. Future research and development should aim at achieving high accuracy, usability, acceptance, and unobtrusiveness. Then, AR could enhance safety and efficiency by improving collaboration in safety-critical operations like ship navigation.

7. ACKNOWLEDGEMENTS

The authors wish to thank Cyriak Heierli for his contributions to the prototype development, the research assistants for the help with the data analysis, Ocean Industries Concept Lab for the design guidance, and the Institute of Marine Research for help in the field study.

8. REFERENCES

1. BATALDEN, B M and SYDNES, A K., "Maritime safety and the ISM code: a study of investigated casualties and incidents," WMU J. Marit. Aff., vol. 13, no. 1, pp. 3–25, Apr. 2014, doi: 10.1007/s13437-013-0051-8.

2. CHAUVIN, C., LARDJANE, S., MOREL, G., CLOSTERMANN, J.P., and LANGARD, B., "Human and organisational factors in maritime accidents: Analysis of collisions at sea using the HFACS," Accid. Anal. Prev., vol. 59, no. 2013, pp. 26–37, Oct. 2013, doi: 10.1016/j.aap.2013.05.006.

3. LEE, J., YOON, W C., and CHUNG, H., "Formal or informal human collaboration approach to maritime safety using FRAM," Cogn. Technol. Work, vol. 22, no. 4, pp. 861–875, Nov. 2020, doi: 10.1007/s10111-019-00606-y.

4. SÁNCHEZ-BEASKOETXEA, J., BASTERRETXEA-IRIBAR, I., SOTÉS, I., and MACHADO, M DE LAS M M., "Human error in marine accidents: Is the crew normally to blame?," Marit. Transp. Res., vol. 2, no. 2021, p. 100016, 2021, doi: 10.1016/j.martra.2021.100016.

5. BEDWELL, W L., WILDMAN, J L., DIAZGRANADOS, D., SALAZAR, M., KRAMER, W S., and SALAS, E., "Collaboration at work: An integrative multilevel conceptualization," Hum. Resour. Manag. Rev., vol. 22, no. 2, pp. 128–145, Jun. 2012, doi: 10.1016/j.hrmr.2011.11.007.

6. ACEJO, I., SAMPSON, H., TURGO, N., ELLIS, N., and TANG, L., "The causes of maritime accidents in the period 2002-2016," Seafarers International Research Centre (SIRC), Cardiff, Wales, 2018. Accessed: Jul. 18, 2023. Online. Available: https://safety4sea.com/wp-content/uploads/2018/12/SIRC-The-causes-of-maritime-accidents-in-the-period-2002-2016-2018_12.pdf

7. HASSEL, M., ASBJØRNSLETT, B E., and HOLE, L P., "Underreporting of maritime accidents to vessel accident databases," Accid. Anal. Prev., vol. 43, no. 6, pp. 2053–2063, Nov. 2011, doi: 10.1016/j.aap.2011.05.027.

8. WINKLER, M., PERLMAN, Y., and WESTREICH, S., "Reporting near-miss safety events: Impacts and decision-making analysis," Saf. Sci., vol. 117, pp. 365–374, Aug. 2019, doi: 10.1016/j.ssci.2019.04.029.

9. ZHANG, W., GOERLANDT, F., KUJALA, P., and WANG, Y., "An advanced method for detecting possible near miss ship collisions from AIS data," Ocean Eng., vol. 124, no. 2016, pp. 141–156, Sep. 2016, doi: 10.1016/j.oceaneng.2016.07.059.

10. UN Trade and Development, "Review of Maritime Transport 2020," United Nations Publications, New York, NY, USA, UNCTAD/RMT/2020, Nov. 2020. Accessed: Jul. 09, 2024. Online. Available: https://unctad.org/publication/review-maritime-transport-2020

11. NILSSON, H., VAN OVERLOOP, J., ALI MEHDI, R., and PÅLSSON, J., "Transnational maritime spatial planning in the North Sea: the shipping context," 2018. Online. Available: https://maritime-spatial-planning.ec.europa.eu/practices/transnational-maritime-spatial-planning-north-sea-shipping-context

12. MARIN, "To2morrow: Orderly Offshore Traffic," MARIN. Accessed: Jul. 02, 2024. Online. Available: https://www.marin.nl/en/news/orderly-offshore-traffic

13. DNV, "Offshore aquaculture: Setting out to feed the world," DNV. Accessed: Jul. 02, 2024. Online. Available: https://www.dnv.com/expert-story/maritime-impact/Offshore-aquaculture-Setting-out-to-feed-the-world

14. SKOPLJAK, N., "BREAKING: Norway Opens Two Offshore Wind Areas, 4.5 GW Available," offshoreWIND.biz, Jul. 12, 2020. Accessed: Jul. 02, 2024. Online. Available: https://www.offshorewind.biz/2020/06/12/breaking-norway-opens-two-offshore-wind-areas-4-5-gw-available/

15. BELLINI, E., "Equinor, Saipem test offshore solar in Norwegian Sea," PV Magazine, Jan. 15, 2021. Accessed: Jul. 02, 2024. Online. Available: https://www.pv-magazine.com/2021/01/15/equinor-saipem-test-offshore-solar-in-norwegian-sea/

16. GU, Y., GOEZ, J C., GUAJARDO, M., and WALLACE, S W., "Autonomous vessels: state of the art and potential opportunities in logistics," Int. Trans. Oper. Res., vol. 28, no. 4, pp. 1706–1739, Jul. 2021, doi: 10.1111/itor.12785.

17. UN Economic Commission for Europe, Climate change impacts and adaptation for transport networks and nodes. 2020. Online. Available: https://unece.org/sites/default/files/2021-01/ECE-TRANS-283e_web.pdf

18. IMO, "IMO's work to cut GHG emissions from ships," International Maritime Organization. Accessed: Jan. 13, 2023. Online. Available: https://www.imo.org/en/MediaCentre/HotTopics/Pages/Cutting-GHG-emissions.aspx

19. Waterborne Technology Platform, Zero-Emission Waterborne Transport. 2021. Accessed: Feb. 02, 2022. Online. Available: https://ec.europa.eu/info/sites/default/files/research_and_innovation/funding/documents/european_partnership_for_zero-emission_waterborne_transport.pdf

20. CHAUVIN, C., CLOSTERMANN, J P., and HOC, J M., "Situation Awareness and the Decision-Making Process in a Dynamic Situation: Avoiding Collisions at Sea," J. Cogn. Eng. Decis. Mak., vol. 2, no. 1, pp. 1–23, Mar. 2008, doi: 10.1518/155534308X284345.

21. ROWEN, A., GRABOWSKI, M., RANCY, J -P., and CRANE, A., "Impacts of Wearable Augmented Reality Displays on operator performance, Situation Awareness, and communication in safety-critical systems," Appl. Ergon., vol. 80, pp. 17–27, Oct. 2019, doi: 10.1016/j.apergo.2019.04.013.

22. ROWEN, A., GRABOWSKI, M., RANCY, J -P., "Moving and improving in safety-critical systems: Impacts of head-mounted displays on operator mobility, performance, and situation awareness," Int. J. Hum.-Comput. Stud., vol. 150, p. 102606, Jun. 2021, doi: 10.1016/j.ijhcs.2021.102606.

23. ROWEN, A., GRABOWSKI, M., RANCY, J -P., "Through the Looking Glass(es): Impacts of Wearable Augmented Reality Displays on Operators in a Safety-Critical System," Ieee Trans. Hum.-Mach. Syst., vol. 49, no. 6, pp. 652–660, Dec. 2019, doi: 10.1109/THMS.2019.2944384.

24. SÆTREVIK, B., and EID, J., "The 'Similarity Index' as an Indicator of Shared Mental Models and Situation Awareness in Field Studies," J. Cogn. Eng. Decis. Mak., vol. 8, no. 2, pp. 119–136, Jun. 2014, doi: 10.1177/1555343413514585.

25. CANNON-BOWERS, J A., SALAS, E., and CONVERSE, S., "Shared mental models in expert team decision making.," Individ. Group Decis. Mak. Curr. Issues, pp. 221–246, 1993.

26. ENDSLEY, M R., "A Systematic Review and Meta-Analysis of Direct Objective Measures of Situation Awareness: A Comparison of SAGAT and SPAM," Hum. Factors J. Hum. Factors Ergon. Soc., vol. 63, no. 1, pp. 124–150, Feb. 2021, doi: 10.1177/0018720819875376.

27. KOCHER, M G., PRAXMARER, M., and SUTTER, M., "Team decision-making," Handb. Labor Hum. Resour. Popul. Econ., pp. 1–25, 2020.

28. READER, T W., "Team Decision Making," in The Wiley Blackwell Handbook of the Psychology of Team Working and Collaborative Processes, 1st ed., E. Salas, R. Rico, and J. Passmore, Eds., Wiley, 2017, pp. 271–296. doi: 10.1002/9781118909997.ch12.

29. WARNER, N., LETSKY, M., and COWEN, M., "Cognitive Model of Team Collaboration: Macro-Cognitive Focus," Proc. Hum. Factors Ergon. Soc. Annu. Meet., vol. 49, no. 3, pp. 269–273, Sep. 2005, doi: 10.1177/154193120504900312.

30. ENDSLEY, M R., "Direct measurement of situation awareness: Validity and use of SAGAT," Situat. Aware. Anal. Meas., vol. 10, pp. 1–21, 2000.

31. ENDSLEY, M R., and JONES, W M., "Situation Awareness Information Dominance & Information Warfare." Defence technical Information Center, Feb. 1997. Accessed: Jul. 09, 2024. Online. Available: https://apps.dtic.mil/sti/citations/ADA347166

32. COOKE, N J., SALAS, E., CANNON-BOWERS, J A., and STOUT, R J., "Measuring Team Knowledge," Hum. Factors J. Hum. Factors Ergon. Soc., vol. 42, no. 1, pp. 151–173, Mar. 2000, doi: 10.1518/001872000779656561.

33. ORASANU, J., "Evaluating team situation awareness through communication," in Experimental analysis and measurement of situation awareness, Daytona Beach, FL, USA: Embry-Riddle University, 2000. Online. Available: https://apps.dtic.mil/sti/tr/pdf/ADA522540.pdf

34. MOHAMMED, S., and DUMVILLE, B C., "Team mental models in a team knowledge framework: expanding theory and measurement across disciplinary boundaries: TEAM MENTAL MODELS IN A TEAM KNOWLEDGE FRAMEWORK," J. Organ. Behav., vol. 22, no. 2, pp. 89–106, Mar. 2001, doi: 10.1002/job.86.

35. SALMON, P M ET AL., "What really is going on? Review of situation awareness models for individuals and teams," Theor. Issues Ergon. Sci., vol. 9, no. 4, pp. 297–323, Jul. 2008, doi: 10.1080/14639220701561775.

36. GRUNER, M., ZOUBIR, M., FRANKE, T., SCHWARZ, B., and HEIDINGER, J., "MariData.HEI," 2022, doi: 10.17605/OSF.IO/CBQKZ.

37. AZUMA, R., BAILLOT, Y., BEHRINGER, R., FEINER, S., JULIER, S., and MACINTYRE, B., "Recent advances in augmented reality," IEEE Comput. Graph. Appl., vol. 21, no. 6, pp. 34–47, Dec. 2001, doi: 10.1109/38.963459.

38. FRYDENBERG, S., AYLWARD, K., NORDBY, K., and EIKENES, J O H., "Development of an Augmented Reality Concept for Icebreaker Assistance and Convoy Operations," J. Mar. Sci. Eng., vol. 9, no. 9, p. 996, Sep. 2021, doi: 10.3390/jmse9090996.

39. OH, J., PARK, S., and KWON, O -S., "Advanced Navigation Aids System based on Augmented Reality," Int. J. E-Navig. Marit. Econ., vol. 5, no. 2016, pp. 21–31, Dec. 2016, doi: 10.1016/j.enavi.2016.12.002.

40. MERINO, L., SCHWARZL, M., KRAUS, M., SEDLMAIR, M., SCHMALSTIEG, D., and WEISKOPF, D., "Evaluating Mixed and Augmented Reality: A Systematic Literature Review (2009-2019)," in 2020 IEEE International Symposium on Mixed and Augmented Reality (ISMAR), Recife/Porto de Galinhas: IEEE, Nov. 2020, pp. 438–451. doi: 10.1109/ISMAR50242.2020.00069.

41. NORDBY, K., FAUSKE, J E., GERNEZ, E., and MALLAM, S., "A User Interface Design Framework for Augmented-Reality-Supported Maritime Navigation," J. Mar. Sci. Eng., vol. 12, no. 3, p. 505, Mar. 2024, doi: 10.3390/jmse12030505.

42. VAN DEN OEVER, F., FJELD, M., and SÆTREVIK, B., "A Systematic Literature Review of Augmented Reality for Maritime Collaboration," Int. J. Human–Computer Interact., vol. 0, no. 0, pp. 1–16, May 2023, doi: https://doi.org/10.1080/10447318.2023.2209838.

43. WICKENS, C D., "Situation Awareness: Review of Mica Endsley's 1995 Articles on Situation Awareness Theory and Measurement," Hum. Factors J. Hum. Factors Ergon. Soc., vol. 50, no. 3, pp. 397–403, Jun. 2008, doi: 10.1518/001872008X288420.

44. FRYDENBERG, S., NORDBY, K., and EIKENES, J O H., "Exploring designs of augmented reality systems for ship bridges in arctic waters," presented at the Human Factors, London, UK: Royal Institute of Naval Architects, Oct. 2018.

45. GERNEZ, E., NORDBY, K., EIKENES, J O., and HAREIDE, O S., "A review of augmented reality applications for ship bridges," Necesse, vol. 5, no. 3, pp. 159–186, Jan. 2020.

46. HOUWELING, K P., MALLAM, S C., VAN DE MERWE, K., and NORDBY, K., "The effects of Augmented Reality on operator Situation Awareness and Head-Down Time," Appl. Ergon., vol. 116, p. 104213, Apr. 2024, doi: 10.1016/j.apergo.2023.104213.

47. FALK, M., SAAGER, M., HARRE, M -C., and FEUERSTACK, S., "Augmented Berthing Support for Maritime Pilots Using a Shore-Based Sensor Infrastructure," presented at the International Conference on Human-Computer Interaction, Springer, 2020, pp. 553–559. Online. Available: https://link.springer.com/chapter/10.1007/978-3-030-60703-6_71

48. LAERA, F., FOGLIA, M M., and FIORENTINO, M., "Augmented reality for sailing: a comparative study of head stabilized vs boat stabilized visualization data for wind and bearing angle," Virtual Real., vol. 28, no. 1, p. 31, Mar. 2024, doi: 10.1007/s10055-023-00936-3.

49. LAERA, F., ET AL., "Augmented reality for maritime navigation data visualisation: a systematic review, issues and perspectives," J. Navig., pp. 1–18, May 2021, doi: 10.1017/S0373463321000412.

50. RAMIREZ-PENA, M., ABAD FRAGA, F J., SALGUERO, J., and BATISTA, M., "Assessing Sustainability in the Shipbuilding Supply Chain 4.0: A Systematic Review," Sustainability, vol. 12, no. 16, p. 6373, Aug. 2020, doi: 10.3390/su12166373.

51. VAN DEN OEVER, F., SÆTREVIK, B., FJELD, M., and NORDBY, K., "A Virtual Reality Experiment shows that Augmented Reality can improve Collaboration in Ship Navigation," Sep. 15, 2023. doi: 10.31234/osf.io/ts5wx.

52. LAERA, F., MANGHISI, V M., EVANGELISTA, A., UVA, A E., FOGLIA, M M., and FIORENTINO, M., "Evaluating an augmented reality interface for sailing navigation: a comparative study with a immersive virtual reality simulator," Virtual Real., vol. 27, no. 2, pp. 929–940, Jun. 2023, doi: 10.1007/s10055-022-00706-7.

53. CARTER, S., MANKOFF, J., KLEMMER, S R., and MATTHEWS, T., "Exiting the Cleanroom: On Ecological Validity and Ubiquitous Computing," Human–Computer Interact., vol. 23, no. 1, pp. 47–99, Feb. 2008, doi: 10.1080/07370020701851086.

54. UNITY, Unity. (2023). Accessed: Apr. 03, 2024. Online. Available: https://unity.com/

55. KHRONOS GROUP, OpenXR. (N.D.). Khronos Group. Online. Available: https://www.khronos.org/openxr/

56. Directorate-General for Research and Innovation, Horizon 2020 Work Programme 2014-2015 General Annex G. 2014. Online. Available: https://ec.europa.eu/research/participants/data/ref/h2020/wp/2014_2015/annexes/h2020-wp1415-annex-g-trl_en.pdf

57. ORTHMANN, B., Sjør: AR Supported Maritime Navigation. (2024). Stockholm, SW. Accessed: Mar. 18, 2024. Microsoft HoloLens 2. Available: https://github.com/thbc/sjoer

58. ROGERS, Y., SHARP, H., and PREECE, J., Interaction Design: beyond human-computer interaction. Indianapolis: John Wiley and Sons, 2023.

59. SCHWABER, K., "SCRUM Development Process," in Business Object Design and Implementation, J. Sutherland, C. Casanave, J. Miller, P. Patel, and G. Hollowell, Eds., London: Springer London, 1997, pp. 117–134. doi: 10.1007/978-1-4471-0947-1_11.

60. ZIMMERMAN, J., FORLIZZI, J., and EVENSON, S., "Research through design as a method for interaction design research in HCI," in Proceedings of the SIGCHI Conference on Human Factors in Computing Systems, San Jose California USA: ACM, Apr. 2007, pp. 493–502. doi: 10.1145/1240624.1240704.

61. ROGERS, C., ET AL., "A universal 3D imaging sensor on a silicon photonics platform," Nature, vol. 590, no. 7845, Feb. 2021, doi: 10.1038/s41586-021-03259-y.

62. EIKENES, J O., "What does an ice breaker do?" Ocean Industries Concept Lab. Accessed: Apr. 14, 2021. Online. Available: https://medium.com/ocean-industries-concept-lab/what-does-an-ice-breaker-do-91e034596639

63. VAN BEEK, A., "Heads Up! Supporting Maritime Navigation using Augmented Reality," University of Bergen, Bergen, Norway, 2022. Accessed: Mar. 25, 2024. Online. Available: https://hdl.handle.net/11250/3001935

64. VAN BEEK, A., Sjør: AR Supported Maritime Navigation. (2022). Bergen, Norway. Accessed: Mar. 18, 2024. Microsoft HoloLens 2. Available: https://github.com/Imable/sjoer

65. HEIERLI, C., Sjør: AR Supported Maritime Navigation. (2022). Bergen, Norway. Accessed: Mar. 18, 2024. Microsoft HoloLens 2. Available: https://github.com/cheierli/sjoer

66. HEIERLI, C., "AR Facilitated Vessel Navigation with HoloLens 2," Eidgenössische Technische Hochschule Zürich, Zürich, Switzerland, 2022.

67. Kystverket, "NAIS," NAIS. Accessed: Jun. 27, 2024. Online. Available: https://nais.kystverket.no/

68. Meowsbox, NetGPS. Accessed: Jul. 09, 2024. Online. Available: https://play.google.com/store/apps/details?id=com.meowsbox.netgps&hl=en&gl=US&pli=1

69. BlueCTRL, Blue Box. (N.D.). BlueCTRL, Ålesund, Norway. Accessed: Jul. 09, 2024. Online. Available: https://bluectrl.io/bluebox

70. VAN SCHAIK, J., "Converting lat/lon coordinates to local coordinates for HoloLens apps," DotNetByExample - The Next Generation. Accessed: Jun. 24, 2024. Online. Available: https://localjoost.github.io/converting-latlon-coordinates-to-local/

71. LEITE, B G., SINOHARA, H T., MARUYAMA, N., and TANNURI, E A., "Maritime navigational assistance by visual augmentation," J. Navig., vol. 75, no. 1, pp. 57–75, Jan. 2022, doi: 10.1017/S0373463321000795.

72. BRAUN V., and CLARKE, V., "Using thematic analysis in psychology," Qual. Res. Psychol., vol. 3, no. 2, pp. 77–101, Jan. 2006, doi: 10.1191/1478088706qp063oa.

73. BROOKE, J., "SUS-A quick and dirty usability scale," in Usability evaluation in industry, 1st ed., Boca Raton, FL, USA: CRC Press, 1996, pp. 4–7. Online. Available: https://www.taylorfrancis.com/chapters/edit/10.1201/9781498710411-35/sus-quick-dirty-usability-scale-john-brooke

74. DAVIS, F D., "A technology acceptance model for empirically testing new end-user information systems: Theory and results," 1985.

75. Lumivero, NVivo. (2023). QSR international, Denver, Colorado, US. Microsoft Windows. Available: https://lumivero.com/products/nvivo/

76. MCHUGH, M L., "Interrater reliability: the kappa statistic," Biochem. Medica, vol. 22, no. 3, pp. 276–282, 2012, doi: 10.11613/BM.2012.031.

77. KIM, H., and GABBARD, J L., "Assessing Distraction Potential of Augmented Reality Head-Up Displays for Vehicle Drivers," Hum. Factors J. Hum. Factors Ergon. Soc., vol. 64, no. 5, pp. 852–865, Aug. 2022, doi: 10.1177/0018720819844845.

78. GIBSON, B S., FOLK, C L., THEEUWES, J., and KINGSTONE, A., Eds., Attentional capture: a special issue of visual cognition, First published in paperback. in Visual cognition Special issue, no. 16,2/3. Hove New York: Routledge, Taylor & Francis Group, 2018.

79. KRISTIĆ, M., ŽUŠKIN, S., BRČIĆ, D., and CAR, M., "Overreliance on ECDIS technology: A challenge for safe navigation," TransNav Int. J. Mar. Navig. Saf. Sea Transp., vol. 15, no. 2, pp. 277–288, 2021.

80. NORDBY, K., MALLAM, S C., and LÜTZHÖFT, M., "Open user interface architecture for digital multivendor ship bridge systems," WMU J. Marit. Aff., vol. 18, no. 2, pp. 297–318, Jun. 2019, doi: 10.1007/s13437-019-00168-w.

81. NORDBY, K., GERNEZ, E., and MALLAM, S., "OpenBridge: designing for consistency across user interfaces in multi-vendor ship bridges," in Proceedings of Ergoship 2019, Haugesund: Western Norway University of Applied Sciences, 2019, pp. 60–68. Online. Available: http://hdl.handle.net/11250/2638416

82. FRYDENBERG, S., EIKENES, J O., and NORDBY, K., "Serendipity in the Field. Facilitating serendipity in design-driven field studies on ship bridges," Des. J., vol. 22, no. sup1, pp. 1899–1912, 2019.

83. GRABOWSKI, M., "Research on Wearable, Immersive Augmented Reality (WIAR) Adoption in Maritime Navigation," J. Navig., vol. 68, no. 3, pp. 453–464, May 2015, doi: 10.1017/S0373463314000873.

84. GRABOWSKI, M., ROWEN, A., and RANCY, J -P., "Evaluation of wearable immersive augmented reality technology in safety-critical systems," Saf. Sci., vol. 103, no. 2018, pp. 23–32, Mar. 2018, doi: 10.1016/j.ssci.2017.11.013.

9. AUTHORS BIOGRAPHY

Floris van den Oever is a Ph.D. researcher in Psychology and Human-Computer Interaction at the University of Bergen, Norway. His work focuses on AR and collaboration. He holds an MSc in Human Factors and Engineering Psychology with honours in design from the University of Twente, The Netherlands.

Bjørn Sætrevik is a professor of general psychology at the University of Bergen, Norway. His work focuses on decision-making, perceived risk, team-processes, and human factors. He is a clinical psychologist (2003) with a Ph.D. in cognitive neuroscience (2007), both from the University of Bergen, Norway.

Abel van Beek is a software developer at Svipper in Troms county, Norway. He holds a BSc in Computer Science (Game Technology) from the University of Utrecht, the Netherlands, and a MSc in Media and Interaction Design from the University of Bergen, Norway, as part of which he developed the first iteration of the Sjør prototype.

Bastian Orthmann is a developer and co-founder of Floating Lab, based in Stockholm, Sweden. He specializes in designing interactive solutions for research, art, and enterprises, with a focus on immersive media and bespoke applications.

Kjetil Nordby is a professor of Interaction and Industrial Design at the Oslo Institute of Design. He is the founder and manager of Ocean Industries Concept Lab and the OpenBridge project. His work has focused on interaction design practice and design of advanced digital workplaces.

Morten Fjeld is a professor of Human-Computer Interaction at the University of Bergen, Norway, and Chalmers, Sweden. His work focuses on tangible user interfaces and founded the t2i Interaction Lab. He holds a dual MSc in applied mathematics from NTNU, Norway, and ENSIMAG, France, and a Ph.D. from ETH, Switzerland.

USING SHIP BRIDGE SIMULATORS FOR TOOL EVALUATION IN AN EARLY DESIGN STAGE

H Diepeveen, MARIN, Netherlands

SUMMARY

Simulators have been integral to the maritime industry since the 1950s, evolving from radar training to complex ship bridge simulators. Through an example, this article describes how ship bridge simulators can be used in evaluating tools in an early design stage. A study was conducted with experienced crew members on a motion-based simulator to assess their ability to recognize preconditions for parametric rolling and their effectiveness in preventing it with and without additional tools. A combination of questionnaires, observations, and interviews was utilized to measure situational awareness and decision-making processes. Key findings highlighted that simulators are effective for early-stage testing and do not require the highest level of fidelity to yield valuable results. This article will not go into the results of the study itself, but rather gives a methodological description of the study, to inspire those considering to utilize simulators in early tool evaluation and concept testing.

1. INTRODUCTION

Simulators are a well-established technology in the maritime industry. The earliest (electronic) human-in-the-loop simulators date back to the 1950s, with the introduction of radar simulation training. This was followed up in the 60s by the invention of ship handling simulators, used for studying crew behaviour and ship and port design [1]. Nowadays, simulators are a multi-purpose tool used in many different fields within the maritime sector. Applications range from training (complex) operations to proof of concept testing and human behavioural studies. This article will aim at discussing how ship bridge simulators can be utilized for tool evaluation in multiple phases of the design cycle. This will be done through discussing one project methodology.

1.2 USE CASE

1.2 (a) Problem Statement

The past decade, the size of deep sea container ships has increased dramatically, with the largest ship having a capacity of 24,346 TEU. Although economically beneficial, there is a growing public and political concern about the environmental risks associated with container shipping. According to the World Shipping Council's 2023 report on container loss at sea, an average of 1566 containers were lost annually between 2008 and 2022 [2]. This is a multifaceted problem caused by factors across the entire container shipping supply chain. Container loss is oftentimes the result of a snowball effect starting from mistakes and time pressure in the stowage, leading up to wrong calculations in the loading computer, which in turn affect the operation during sailing. The current study will zoom into the operation.

1.2 (b) Project Scope

A review of incidents performed by MARIN showed that parametric rolling is one of the contributors to container loss at sea. In simplified terms, parametric rolling is a phenomenon that can occur when the wave encounter period is approximately half of the ship's natural period, in head seas, following seas or quartering seas, leading to severe rolling within a short time period. When the preconditions are met, parametric rolling may build up any time and roll angles can increase rapidly to extreme angles, with only a couple of oscillations between mild and extreme angles. What makes parametric rolling complex is that it is rather difficult to predict if and when it will occur, while it is highly important to take preventative actions. Acting upon it while angles are already being exceeded, would not allow for sufficient time to make course and speed changes. Therefore, it is important for a container ship crew to be aware of parametric rolling preconditions, have the appropriate tools and knowledge to recognise it, and have an understanding on how to take preventative actions.

To further investigate this topic, the current study focussed on the following research questions:

- Can a bridge crew recognise preconditions of parametric rolling?
- What actions does a bridge crew take to prevent parametric rolling and are they effective?
- Does adding additional tools increase the situational awareness and decision-making of the bridge crew in recognising and preventing rolling?

This article will discuss the methodology that was applied to answer these questions and the upsides and downsides of the methodology.

2. METHODLOGY

2.1 PARTICIPANTS

Participants were recruited using a snowball sampling method with the following requirements:

- Captain, first officer, second officer or third mate (making navigational decisions);
- Recent experience on a 11 to 24 kTEU container ship;
- Varying levels of experience;
- Communicating in English.

Eleven of the participants met all of the requirements. Three participants did not meet the container ship size requirement, but had sufficient experience to offer valuable input for the study. Figure 1 shows an overview of the years of sailing experience of participants.

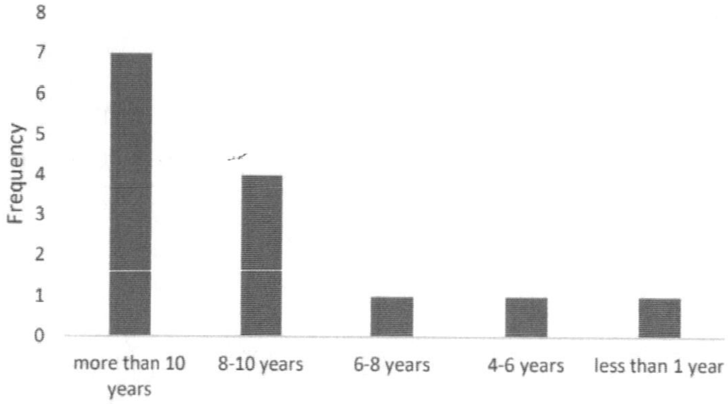

Figure 1. Years of sailing experience of participant sample.

2.2 SIMULATOR SET-UP

The experiment was done on MARIN's Large Motion Simulator (LMS): a six degrees of freedom motion-based bridge of 4 x 5 m on a hexapod with a payload of 14,000 kg and 360 degrees projection in a large spherical dome with a diameter of 16 m. Figure 2 illustrates what the simulator looks like.

The displays on the bridge provided the following information:

- A conning display showing the forward speed, the lateral speed fore and aft, the rate of turn, the setting of the rudder, keel clearance, wind speed and direction;
- Double Kelvin Hughes radar;
- Electronic Nautical Chart, showing the position and the predicted path of the vessel;
- VHF equipment for communication with the instructor;
- Overhead displays with STW/SOG, propeller revolutions, ROT, heading repeater, true and relative wind indicators and keel clearance;
- Two consoles in which additional information is shown (will be further elaborated in section 2.3)

The experimental runs were selected based on model tests conducted at MARIN with a 15kTEU container vessel. 15-minute snippets of the selected runs were replayed on the LMS, to allow participants to experience preconditions of parametric rolling. To make the runs more realistic and to offer a distraction, traffic was also introduced in the runs.

Figure 2. Large Motion Simulator (LMS).

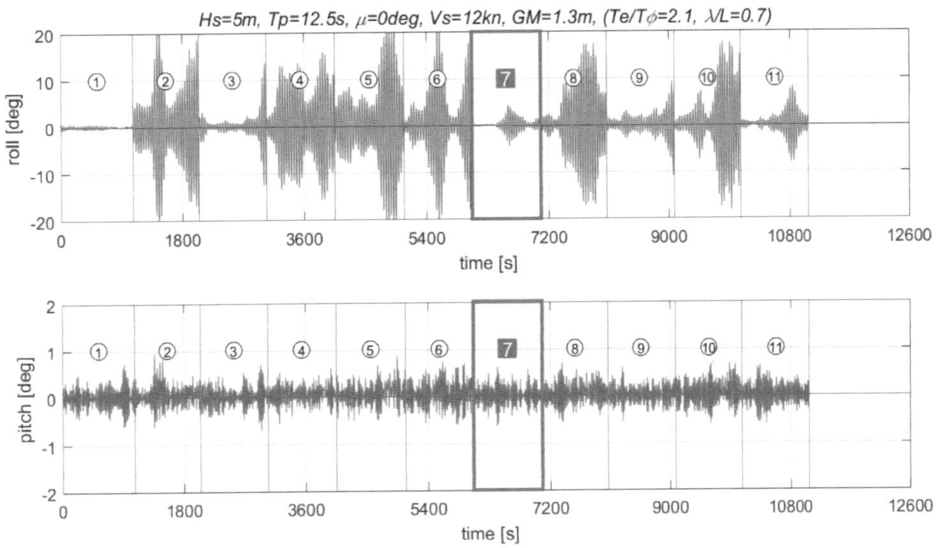

Figure 3. Illustration of model test time trace of parametric rolling in following seas. The snippet used in the simulations is marked in red.

2.3 EXPERIMENTAL CONDITIONS

2.3 (a) Experimental Runs

The participants experienced a total of six experimental runs and one familiarization run, numbered with the letters A to F. The run type was counterbalanced across participants, meaning that each participant or participant duo experienced the runs in a different order. Table 1 gives an overview of the runs. The column "max. roll experienced" describes the maximum roll that was actually experienced by the participants in the 15-minute run. The column "max. roll in time trace" describes the maximum roll participants would have experienced if the entire model test had been replayed. In this study, we were explicitly interested in the recognition of and reaction to preconditions of parametric rolling, because actions when heavy rolling is already taking place are oftentimes too late. We therefore chose snippets in the time traces that occurred right before heavier parametric rolling took place.

Run	Ship Behaviour	Max. roll experienced	Max. roll in time trace
A	Parametric roll in bow quartering seas	2.5	3.1
B	Resonant roll	9.1	10.1
C	Parametric roll in following seas	4.6	23.9
D	Parametric roll in stern quartering seas	8.4	22.4
E	Parametric roll in stern quartering seas	15.0	18.7
F	Parametric roll in stern quartering seas	9.7	17.1

Table 1. Overview of experimental runs.

2.3 (b) Information

Each participant experienced three types of information throughout the runs, building up from basic to more advanced information. The information types are numbered I to III and can be found in Table 2. The familiarization run and first run were always done with access to information condition I (basic), the middle runs were always done with access to information condition II (advanced) and the last run was always done with information condition III (advanced_extra). The polar plot/roll risk estimator was offered in a static form on screens on both port and starboard side. Figure 4 shows an example of what this screen looked like. The electronic roll/pitch graph showed live data on the roll and pitch motions of the ship over time and was displayed on the ECDIS.

Information	I: basic	II: advanced	III: advanced_extra
Outside world view	X	X	X
Conning	X	X	X
Radar	X	X	X
Electronic Nautical Chart	X	X	X
VHF equipment	X	X	X
Overhead displays	X	X	X
Inclinometer	X	X	X
Weather forecast	X	X	X
Polar plot with roll risk estimation		X	X
Electronic roll/pitch graph			X

Table 2. Overview of information.

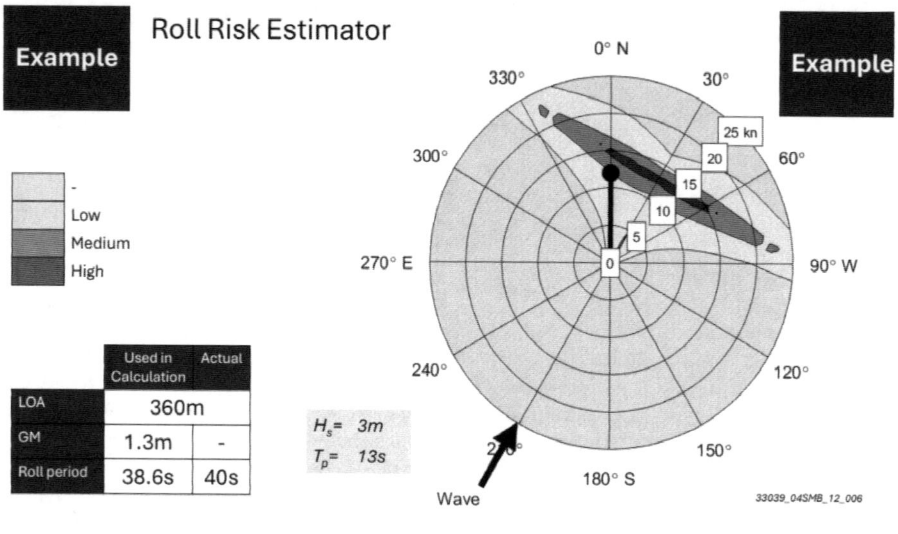

Figure 4. Example of Roll Risk Estimator.

2.4 THE TASK

The participants were not informed that they would be experiencing parametric rolling or that the scope of the experiment was to react to parametric rolling behavior. This decision was made to collect data that was as unbiased as possible. The participant was therefore given the following main task "do what you would normally do during watchkeeping".

Beside the main task, participants were also given experiment specific tasks:

- Raise your hand when a situation that occurs requires your attention or you would start preparing for an action;
- Communicate navigational actions to the helmsman;
- Also communicate any non-navigational action you would take

2.5 THE MEASURES

2.5 (a) Situational Awareness (SA)

A Situational Awareness questionnaire was created based on Endsley's situational awareness model [3], shown in Figure 5. The Endsley model breaks situational awareness up into three levels:

- Level 1: the operator becomes aware of their environment, perceiving all information available.
- Level 2: the operator tries to understand the available information in the context of the environment.
- Level 3: the operator anticipates on the situation, predicting future events.

When the operator has gone through all three levels of situational awareness, a decision can be taken on actions.

The questionnaire chronological addresses all three levels of situational awareness on a 5-point Likert scale, in which 1 represents strongly disagree and 5 represents strongly agree. Furthermore, it also contains one open question about decision making and one closed question about surprise. The latter was inspired by the first question on the Situation Awareness Rating Technique (SART) [4] and was chosen to measure the perceived predictability of events. An overview of the questions is given in Table 3.

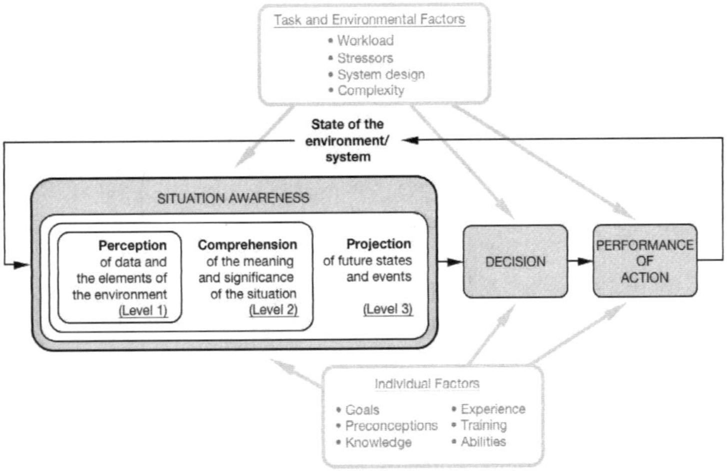

Figure 5. Endsley model of situational awareness.

Question	Type
I had enough information available to perceive what was happening in my environment.	5-point Likert (strongly disagree – strongly agree)
I understood the available information.	5-point Likert (strongly disagree – strongly agree)
The information that was available to me helped me in anticipating events.	5-point Likert (strongly disagree – strongly agree)
I used all the available information to decide whether or not to take action.	5-point Likert (strongly disagree – strongly agree)

Which information did you use in your decision making? Please also mention it if you used any outside information or prior knowledge and experience in your decision making.	Open
At any moment in the simulation, I felt surprised about what happened.	5-point Likert (never – very often)

Table 3. Overview of situational awareness questions.

2.5 (b) Workload

Workload was assessed making use of the Rating Scale Mental Effort (RSME) [5], alongside an additional question about level of attention. The RSME is quick 150-point scale, used for mental load assessment. The level of attention question was asked on a 10-point scale and was assessed based on the premise that a higher required level of attention correlates with a higher mental load.

2.5 (c) Observations

Observations were done in the control room of the Large Motion Simulator, as shown in Figure 6. In the control room, the researchers had the ability to see and hear the participants through video and audio devices. Through the observations, data was collected on the actions that were taken by the participants, the information that was being used and the timing of the actions. These actions were later drawn into the polar plots and roll/pitch graphs to compare the type of actions, the timing of actions and assess the efficiency of actions.

Figure 6. Control room of LMS.

2.5 (c) Interviews

After the runs and during the debriefing, participants were further asked about their decision making and actions taken through semi-structured interviews. In the debriefing, participants were informed about what they had experienced through explanations of parametric rolling and illustrations of each run (see Figure 3). Below, a selection of the questions that were asked are listed:

1. What was for you the reason to take action?
2. What information played a role in the decision to take action?
3. Could you further explain why you took action now?

4. Did you expect that the situation would have unfolded itself like this, if no action had been taken? (in debriefing)
5. Would you have taken a different action knowing what you know now? (in debriefing)
6. Did you ever experience parametric rolling in your sailing career? (in debriefing)
7. Do you feel like you could recognize the preconditions of parametric rolling in the runs? (in debriefing)

3. CONCLUSIONS

Upsides of this methodology:

- Evaluating a system or tool on a simulator is an efficient way of gaining an understanding of an operator's situational awareness and decision-making process. It allows a researcher to observe the system or tool in the context at which it will be used, helping the researcher understand:
    - When the tool is being used (e.g. in which phase of the OODA loop)
    - Why the tool is being used
    - How the tool is being used

  An example of a finding giving input to the first bullet point is given in Table 4.
- Simulators can already be used in an early stage of concept testing. In the present study, one of the tools that was offered to the participants was merely a static image. We gained however a lot of feedback and understanding about the tool, because it allowed enough information for the participants to use and spoke enough to the imagination to give feedback on it.
- Not every aspect of the simulation has to be worked out to the highest fidelity level. In the present study, participants were not capable of controlling the ship. This did not limit the sense of immersion participants experienced, neither did it hamper the quality of the results that were obtained. This is well reflected in Figure 7.
- Studying situational awareness through a combination of questionnaires, observations and interviews gives a more complete image of situational awareness and helps in filling out missing pieces. One result of the current study was that participants evaluated their situational awareness high on the questionnaire throughout conditions, while generally showing poor situational awareness during observations and through interview questions. This helped in gaining the insight that participants were often unaware of their own misunderstanding of the situation, highlighting the difficulty of recognizing parametric rolling and the need for clear and transparent support.

OODA	Decision making pattern 1	Decision making pattern 2
Observe	The participant looks at all information available to him/her to reach level 1 situational awareness.	The participant looks at all information available to him/her to reach level 1 situational awareness.
Orient	The participant tries to get an understanding of the available information. The participant sees the roll risk in the polar plot and addresses to be more vigilant. The participant proceeds to observe the ship behaviour through the outside view, inclinometer, and their own feeling.	The participant tries to get an understanding of the available information. The participant sees the roll risk in the polar plot and addresses the desire to avoid these conditions.
Decide	The participant observes that the roll angle of the ship is increasing. The participant looks at the polar plot to see what course and/or speed changes would bring the ship into safer conditions.	The participant decides to take action right after seeing the polar plot to avoid unsafe roll angles.
Act	The participant changes course and/or speed.	The participant changes course and/or speed.

Table 4. Example of one of the results of the study, giving input about decision making when using information.

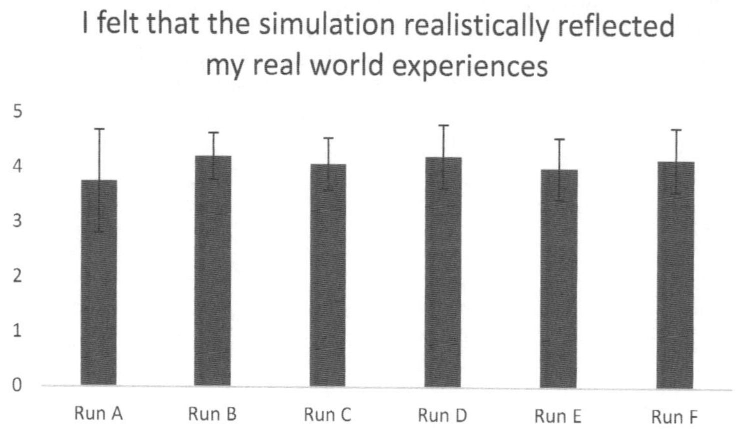

Figure 7. Sense of presence of participants.

Downsides of this methodology:

- Although simulator studies can vary in level of fidelity, they are more time consuming and expensive to set up than some other methodologies for early tool evaluation and concept testing.
- Relating to the previous point, it is difficult to do an assessment with a large participant sample, due to time constraint. It should however be noted that it is pretty common within UX research to do tool assessments with a small participant group.
- In this particular study, it was quite challenging to eradicate bias completely, because participants gradually get more understanding about the scope of the study throughout runs. Especially when more scenario specific tools are being introduced. This can be partly resolved through counterbalancing, but is difficult to fully eradicate.

4. REFERENCES

1. *Simulators: 45 years of development and debate*. (October, 2021.). https://www.nautilusint.org/en/news-insight/telegraph/simulators-45-years-of-development-and-debate/

2. World Shipping Council. (May, 2023). *Containers Lost at Sea – 2023 update*

3. ENDSLEY, M R., (1995). Toward a Theory of Situation Awareness in Dynamic Systems. *Human Factors The Journal Of The Human Factors And Ergonomics Society*, *37*(1), 32–64. https://doi.org/10.1518/001872095779049543

4. TAYLOR, R M., (1990). Situational Awareness Rating Technique (SART): The development of a toll for aircrew systems design. In: *AGARD Conference Proceedings No 478, Situational Awareness in Aerospace Operations. Aerospace Medical Panel Symposium*, Copenhagen, 2nd -6th October 1989.

5. ZIJLSTRA, F. & DOORN, L., (1985). The Construction of a Scale to Measure Perceived Effort. *Department of Philosophy and Social Sciences*.

5. AUTHOR'S BIOGRAPHY

Heike Diepeveen is a Human Factors Specialist working at MARIN. Heike has a background in Applied Cognitive Psychology and is responsible for a diverse range of human factors projects within MARIN. Common topics she works on are HMI evaluation and design, concept development and the application of virtual reality within a maritime context.

"ANY FOOL COULD TELL HOW BAD THE SHIP WAS": THE IMPORTANCE OF EMPLOYEE VOICE IN SAFETY CULTURE

I Bron, Toronto Metropolitan University, Centre for Free Expression, Canada
J Dalziel and **R Pelot**, Dalhousie University, Canada

SUMMARY

After a disaster, many questions are raised by the public about how the incident could possibly have happened. Weren't there procedures, standards, rules, and laws in place to prevent this? Weren't there safety inspectors who were obligated to protect the public safety? Standards, procedures, rules and laws exist to maintain public safety in our increasingly complex world. But, as many maritime disasters and other incidents painfully demonstrate, they count for little unless there is a Safety Culture to ensure accountability, and that they are applied and enforced. The authors explore the pressures on the members of the safety network to "look the other way" and allow non-compliant and unsafe practices to continue. The pressures on those who try to maintain standards and protect the public safety are explored; the failure of their co-workers and society in general to support them is exposed. Solutions to these issues are proposed.

1. INTRODUCTION

No organization or industry is immune to error, mismanagement, and corruption. Detecting and correcting such instances is, and may always be, a game of cat and mouse involving regulators, corporate executives and boards, and wrongdoers. In the marine sector, the International Maritime Organization (IMO), the International Labour Organization (ILO), National Administrations and private regulatory organizations provide a regulatory regime which, if followed, substantially reduces the risk and severity of maritime casualties.

One aspect of this regime is the International Safety Management (ISM) Code, implemented in 1993 as a response to the loss of the *Herald of Free Enterprise* off Zeebrugge in 1987, in which both design and management flaws played a role. Members of the ship's crew had raised concerns but were ignored. To a degree, the ISM Code addresses this failure by requiring that all operators institute a safety management system (SMS) which must include clear channels for communication (s. 1.4.3) and for procedures for disclosure of accidents and noncompliance with the Code (s. 1.4.4). Other sections support these requirements. While the Code never uses the term, this suggests whistleblowing avenues. The requirements of the ISM Code have also been enshrined in national laws, such as in U.S. 33 CFR 96.100, the U.K. Port Marine Safety Code and Canada's Safety Management Regulations.

Despite the existing and explicit requirements, noncompliance and failure continues – sometimes with fatal consequences. These cases often feature people who were aware of the problem, reported it, and were either ignored or punished. Even captains are not immune. In other cases, employees opt to remain silent. This is unfortunate, as whistleblowing is an efficient and effective way to detect misconduct and error, benefiting the company and the public interest.

In this paper, the authors explore this complex problem using concrete examples and propose possible solutions. The paper is based on the authors' research and personal experiences, including whistleblowing when attempting to maintain safety standards. It begins with a definition of terms, then examines cases where employee voice could have prevented catastrophe and concludes with a description of best and worst practices for disclosures.

2. THE DEFINITION AND VALUE OF SPEAK-UP AND WHISTLEBLOWING

Whistleblowing is a contentious topic, burdened with misconceptions, cultural and organizational norms on loyalty, and ignorance of the benefits of robust speak-up arrangements. Some benefits are obvious, as in the case of the *Herald of Free Enterprise*, where acting on requests for restrictions on passenger capacity and a warning light might have saved 193 lives [1, Paras. 17–18]. Others are more subtle, such as improved organizational performance.

2.1 DEFINITION

Whistleblowing has various definitions, but the most used is "the disclosure by organization members (former or current) of illegal, immoral or illegitimate practices under the control of their employers, to persons or organizations that may be able to effect action" [2, p. 4]. "Action" includes the power to investigate, correct, or publicize the wrongdoing. Wrongdoing includes non-trivial actions, processes, or omissions.

Whistleblowing is a sub-set of speak-up culture. Speaking up is broader, including expressing unpopular opinions about existing processes or results, challenging unspoken assumptions, or other forms of dissent. It challenges the status quo and thus may carry some risk, such as damaged relationships, being treated poorly or humiliated, or formal job action [3]. Whistleblowing, on the other hand, is more specific and riskier because it identifies misconduct and directly implicates colleagues or superiors. For this reason, it has been associated with disloyalty. Attitudes shifted as scandals and preventable disasters revealed that insiders had known of problems but had been ignored or punished – or had not dared to speak up. Later, Vandekerckhove and Commers [4] reframed whistleblowing as "rational loyalty" – that is, loyal to the stated, legitimate goals of organizations, and not to persons or practices therein. For example, placing loyalty to a hospital's administration before patient care would be considered aberrant by society and a serious breach of professional duties [5].

Perhaps contributing to lingering negative attitudes, it is a popular misconception that whistleblowing is limited to disclosures that become public record – that is, external disclosures to the media or a regulator. However, rather than being a discrete act, whistleblowing is a process: Nearly all (98%) external whistleblowers first attempt internal channels before taking their concerns outside the organization [6]. There is little evidence that organizations view internal disclosures as loyal; internal whistleblowers are treated much the same as external whistleblowers [7], [8].

Expectations that whistleblowers be morally spotless are also pernicious, as whistleblowers are human and may thus have ulterior motives, such as vindication and revenge against those making reprisals. Given the difficulty in establishing motive, the consensus is that it is the information that matters e.g., [9, p. 500]. Where the information is true, or if the reporter has a reasonable belief that it is true, the disclosure of wrongdoing is whistleblowing. Deliberate fabrications are not whistleblowing.

2.2 THE PLACE OF SPEAK-UP AND WHISTLEBLOWING IN MARITIME INDUSTRIES

Speak-up and whistleblowing are essential components of a sound safety culture. Safety culture is characterized by the actions and intentions of organizational members being oriented towards the prevention of accidents and the minimizing of occupational hazards. Reason [10] proposes that it comprises several elements, including being an informed culture, a reporting culture, a learning culture, a just culture, and a flexible culture. It requires depth and commitment from leadership to the most junior worker, maximizing the likelihood that an error by one individual will be detected and corrected by others.

An SMS is intended to institutionalize safety culture. In the maritime environment, it should provide a link between onboard safety management and a designated person ashore (DPA) who has access to the highest level of management in the company. The designated person is responsible for monitoring the safety of the ship and ensuring that adequate resources and shore-based support are provided. The system places responsibility on the master and the designated person to communicate information related to onboard safety deficiencies and non-conformities and their possible causes, regulations pertaining to corrective actions, and recordkeeping of such actions. It also calls for the creation of a documented record of deficiencies that could prove negligence. The intent is to make it impossible for the company to plead ignorance of defects and ensure that shore-side management is held directly responsible and liable for the safe condition and operation of the ship. Essential to this are open avenues of communication and procedures for employees and other staff to make reports safely. Where there is any risk of reprisal after speaking up, the chances of reporting diminish [7] – undermining the SMS.

It should be noted here that a threat to all SMS is regulatory capture. Regulatory capture occurs when industry, or a portion of an industry, gains control over the regulatory body responsible for its oversight. This can occur in several ways, such as "revolving door" personnel transfers between the regulator and the private sector, excessive influence on politicians or senior government bureaucrats, or simple bribery. Once complete, the regulator cooperates with industry actors to subvert or ignore regulations. This phenomenon is not limited to developing countries. Investigations after the Deepwater Horizon explosion in the Gulf of Mexico found that the Minerals Management Service, was much too close to the oil industry so, even when workers expressed concerns, nothing was done [11].

2.3 BENEFITS TO ORGANIZATIONAL PERFORMANCE

Besides compliance with ISM Code, there are other immediate and long-term benefits to effective speak-up and whistleblowing arrangements. Semi-annual surveys by the Association of Certified Fraud Examiners [12] have consistently found that internal reporting is the most effective means of detecting fraud (42%), surpassing the next closest method (audit, 16%). In this way, it serves as a form of decentralized enforcement that reduces the costs of detecting wrongdoing [13]. This, in turn, facilitates its early detection and correction [14]. Reputation is also protected, as whistleblowers will usually try to disclose internally but will take matters to the public if the response is unsatisfactory or if the climate appears hostile [6], [15, p. 1284].

In contrast, not having an effective disclosure system increases the risk that false or mistaken disclosures consume time needlessly and cause reputational damage. This occurred in Canada, when an employee reported false (but honestly believed) accusations about the handling of personal data at BC Health. An amateurish, botched investigation led to false accusations, erroneous findings, a suicide, and contributed to the fall of the government [16].

In the longer term, the mere existence of an effective whistleblowing regime can have a deterrence effect on misconduct [17]. Speak-up culture has been found to increase overall efficacy and efficiency where employees work cooperatively [18]; similar findings have been obtained for ethics and compliance programs, which also require reporting channels [19]. Other potential benefits for the organization include aiding recovery of lost assets, ensuring compliance with organizational policies, procedures, and legal and social obligations, and attracting and retaining personnel committed to the organization's values and culture [20].

For these reasons and others, whistleblowing systems are now advocated as an important part of ethics frameworks, risk management, and good governance, as well as a tool to fight corruption and fraud, enhance efficiency, and regain some social control over large, complex organizations [14], [21]. As noted by one author, "All staff members are, and should be, potential whistleblowers" [22].

3. REPRESENTATIVE CASES WHERE EMPLOYEE VOICE MATTERED

There is persistent evidence that public interest disclosure (whistleblowing) is a valuable mechanism for countering malfeasance, improving companies' practices, protecting the public and enabling companies to resolve issues before they become worse. This is best demonstrated through actual cases of disasters that could have been preventable if whistleblowers were not intimidated or their warnings were heeded. The following sections highlight a variety of such cases in the shipping industry, and some in other domains that illustrate the ongoing wrongdoings, often leading to the unnecessary loss of lives.

3.1 THE *PRINCESS ASHIKA*

We adopt, as an accurate opinion and assessment of MV *Princess Ashika*, the remarks made by a Fijian Maritime Safety Authority surveyor involved in the annual survey of the vessel on 16th December 2008, where he said: "The vessel is now 36 years old and the condition it is now in if it is allowed to trade will cause a maritime disaster and will be the cause of pollution. It is about time the vessel be condemned for good since its condition will continue to deteriorate further. The vessel is beyond repair and is no longer fit for sea services of any nature. We cannot ignore the fact that she is truly unseaworthy. [23]

Subsequent to the foregoing survey report, the *Princess Ashika* was sold to Tonga and completed four voyages. The ship survived the fourth voyage because a crew member kicked a hole in the side of the steel-hulled vessel to let water off the car deck. On the fifth voyage, in August 2009, she sank, with the loss of 74 lives. She had been issued the necessary certificates prior to her fateful voyage [24]. However, senior officers of the ship subsequently told police that they knew the 37-year-old ship was unsuitable, unsafe and unseaworthy – but accepted the ship's safety certificate anyway.

The master said he knew the ship was old and corroded but did not resign as it was his only way to make a living. The man who surveyed the MV *Princess Ashika* before it left Tonga on its fateful last trip says he was afraid he would lose his job if he told a government minister about the dangerous condition of the ferry [25]. The Shipping Corporation of Polynesia argued that it relied on certificates "issued by the appropriate authority", and accordingly operated *Princess Ashika* lawfully. However, the evidence as to the unseaworthiness and appalling state of the vessel was overwhelming and compelling.

Mr. Sione Mafi Kavaliku, a marine officer in the employ of the Ministry of Transport, when asked by the Acting Director of Marine and Ports as to his opinion about the vessel, aptly responded by advising that "any fool (could) tell how bad the ship was" [26]. Tonga established the Royal Commission of Inquiry into the Sinking of the *Princess Ashika*. It stated: "There were many causes of the disaster. The tragedy is that they were all easily preventable and the deaths were completely senseless. It was scandalous that such a maritime disaster could ever have been allowed to occur. It was a result of systemic and individual failures" [23].

Figure 1: Photograph taken on 14 July 2009 showing holes caused by waves during the second voyage. Patched with a doubler [23]

3.2 THE *HOLIDAY ISLAND*

"*Transport Canada documents say extensive corrosion should have been visible before 2016.*" [27]

The 99-meter ferry *Holiday Island* was built in 1971 for seasonal service (May to October) across the Northumberland Strait to Prince Edward Island. Starting in 2007, she was inspected annually under the Delegated Statutory Inspection Program (in which a classification society performs inspections on behalf of the flag administration, Transport Canada Marine Safety). The vessel was inspected annually and found to be in satisfactory condition and certified as complying with class and flag state requirements. Despite a May 2015 inspection report that stated all tanks had been inspected and that their structural condition was good, in 2016 an inspection revealed many serious, obvious structural deficiencies such as extensive diminution, tank bulkheads perforated by corrosion, and the main underdeck structure corroded and fractured. The ship was pulled out of service for almost the entire season for extensive steel repairs.

Transport Canada officials noted that "It would appear reasonable to expect that some of the deficiencies noted where perforation and excessive corrosion is present and requires immediate repair ought to have been observable previous to May 2016 based on previous inspection." Further, Transport Canada officials required that Lloyd's Register "provide an explanation of the apparent irregularity concerning the inspection" of the tanks. Transport Canada officials claimed to have monitored Lloyd's inspection work on the ferry in 2014 and 2016, and that after the 2016 suspension of *Holiday Island* they performed an audit of Lloyd's Register. An action plan was developed to address the findings, and a follow-up was conducted by Transport Canada [27].

Many parties should have been aware of the decay of the vessel and did not report or address them. Without adequate whistleblowing opportunities and protections, however, there is little incentive to bring such serious deficiencies to light. Of course, ship inspection involves a lot more than just steel; it includes the condition of the machinery and electrical systems, fire safety, and many other factors. A month after Transport Canada's concerns were published by the CBC [27], the *Holiday Island* caught fire. Fortunately, it was fine weather, midday, near port and with only a partial load of passengers, so over 200 people were able to evacuate without serious injury. The fire continued to smoulder in the engine room for over 24 hours before being extinguished, suggesting that the fire suppression systems did not work as intended. The *Holiday Island* was taken out of service and scrapped. Despite this, the operator stated: "At no time, before or after 2016, was the safety of the vessel... at risk" [27].

Figure 2: The MV *Holiday Island* fire forced the evacuation of 200 people [28]

3.3 THE *OCEAN RANGER*

In February 1982 the oil drilling rig *Ocean Ranger* was lost with all hands, 84 souls, off the coast of Newfoundland. The loss was the result of several factors, including a design which allowed water to enter through portlights in the ballast control room, affecting the electronic systems there, poor training in manual control of ballast systems, inadequate lifesaving gear, a toxic work culture in which the captain could be overruled by a rig's top oil "pusher," and a completely inadequate regulatory framework [29].

3.4 THE *MARINE ELECTRIC* (1983) AND *EL FARO* (2015)

In February 1983 the *Marine Electric*, a 'reconditioned' World War II vessel, was on a routine voyage thirty miles off the East Coast of the United States when the old coal carrier sank, 31 crewmembers died, three survived. Like many other American ships, the *Marine Electric* was very old and in poor condition. Yet the American authorities, and American Bureau of Shipping (ABS) inspecting on their behalf under the existing alternate compliance program (ACP) persistently turned a blind eye to the potential dangers.

A subsequent Coast Guard formal inquiry led to a criminal indictment of *Marine Electric*'s owners. Despite been recently inspected by both ABS and Coast Guard inspectors, serious deficiencies had been overlooked, the deck and hatch covers most likely allowed flooding of the forward compartments. Subsequently, greater safety standards were introduced; about 70 fully certified vessels were scrapped [30].

Three decades later, the lessons of *Marine Electric* were largely forgotten. Commercial and financial priorities took precedence over technical standards and safety once again. As a warning to others, whistleblowers were shown the door, others kept their head down. On October 1, 2015, the large American roll-on/roll off container ship *El Faro* sailed directly into Hurricane Joaquin and was lost with all hands [31].

Figure 3: Danielle Randolph, one of 33 lost from *El Faro* [32]

Why did Captain Davis ignore the warnings, by the weather services, by his junior officers, that the hurricane had changed its track, and he was heading straight into it? Was it that some years previously he had lost his position as captain because he was careful, and someone in the office thought he was too careful? Once bitten, twice shy.

> *It is difficult to get a man to understand something, when his salary depends on his not understanding it.* (Upton Sinclair)

Notwithstanding the master's decision, why did the 30,000-ton ship not survive the hurricane? It had recently passed both class and flag state inspections. The subsequent investigation revealed that the *El Faro* had many deficiencies which allowed the ship to flood, list, lose power and eventually capsize. The Coast Guard Commandant stated:

This is a call to action for the entire maritime community. TOTE, ABS, and the Coast Guard must learn and move with a sense of urgency. This tragic story points to the need for a strong and enduring commitment to all elements of the safety framework. First and foremost, the company must commit to safety culture by embracing their responsibilities under the ISM Code. Secondly, recognized organizations must fully and effectively perform their duties and responsibilities. Finally, the Coast Guard must, and will, provide the final safety net with sustainable policy, oversight, and accountability. [33]

Subsequently, six fully certified ships were scrapped outright, and another 50 (roughly half of the U.S. flag merchant fleet in 2018) were flagged for high-priority inspections. Monitoring the performance of the recognized organizations (ROs) under the ACP increased [31]. It is noteworthy that the *Marine Electric* sinking took place in 1983, ten years before the introduction of the ISM Code; the *El Faro* sinking in 2015 – 23 years after.

3.5 SPOTLIGHT ON SAFETY: WHY ACCIDENTS ARE OFTEN NOT ACCIDENTAL

The *Spotlight* paper relates the cases of two American ship captains fired for reporting safety violations to shore-side management and the Coast Guard. Appendix B notes that Canadian safety inspectors were unable to do their job [34]. After the intervention of their union, a large ferry was scrapped, another vessel had extensive repairs.

The *Spotlight* paper makes a series of seven proposals (pp. 10-11) to improve safety and protect ship's crews.

The regulatory regime exists, the reporting requirements are defined in the ISM Code, however the report notes that "It is difficult to establish a shared safety culture between the ship and shore-side management when the future careers of the master and crew may depend on not sharing the safety information with management that they are required to share by law" [34].

It was subsequently entered into the record of the United States Congress.

Figure 4: Poster Presentation – World Maritime Rescue Conference, Vancouver, 2019 [34]

3.6 THE *BARRACUDINA* – A PERSONAL ACCOUNT

On November 25, 1978, Captain Jim Chaulk, Master of the fishing vessel *Barracudina*, radioed he was 20 miles out and would be in Port aux Basques, Newfoundland, in two hours. That short communication was the last ever heard from the *Barracudina* and her crew of five.

Figure 5: Memorial for the crew of the FV *Barracudina* [35]

John Dalziel, a professional engineer, had supervised the construction of the *Barracudina*. He had also sailed with her and knew Captain Chaulk and his crew personally [35]. This experience marked him deeply. He was later involved in the construction, refit and repair of other vessels and spent time at sea. Through this, he endeavoured to remember how important it was that he fulfil his professional responsibilities with integrity, even though at times there may be pressure to do otherwise.

Twenty years later, as a marine surveyor in the Halifax office of a classification society, he was ordered to falsify reporting on a bulk carrier about to make a winter trans-Pacific voyage to China for special survey repairs, and to delete safety concerns from the ship's file. This included issues such as numerous leaks in the tank top and plywood installed in ballast tanks to prevent personnel from falling through internal structure. Diminution of ship structure exceeded class allowances to the point that the thickness measurement technician had knocked holes through the engine room bulkhead.

Despite this, other surveyors had reported that this was satisfactory. The superintendent expressed serious concerns to him about the safety of the voyage, based on about 800 tons of steel needing renewal. He was told that class had sided with the owner so that the ship could sail to China, presumably to save money, and later learned that about 2,800 tons of steel were eventually cropped and renewed, more than a quarter of the ship's total.

When he refused to falsify records, he reported this internally and then was constructively dismissed. He complied with the Professional Engineers Code of Ethics and reported to his professional society. The society supported his complaint, concluding that his employer had acted improperly. Despite this, it did not take any other action. His subsequent claim under the *Canada Labour Code* was met with protests by his former employer that the Canadian federal government had no jurisdiction, in that it was not an "essential part of the shipping" (in Canada shipping being under the Federal jurisdiction). This lengthened the process by years as it went through three levels of the legal system. It was also raised in Parliament and reported internationally. Ultimately, the Federal Court rejected his employer's arguments, and the case was eventually settled. The reasons, he argues, were that his former employer "had run out of delaying tactics, and my lawyer and I had run out of money." He believes his case did send a clear signal to professional colleagues – do not challenge management.

In later years, he held several roles with the flag administration, up to Acting National Manager of Small Vessels at Transport Canada. He, safety inspectors, and senior regional management raised concerns about the effectiveness of Canadian safety inspection programs, including Transport Canada's "partners in safety" programs. Their concerns fell on deaf ears or were met with pressure to remain silent. He believes that he and like-minded colleagues were passed over for promotion as a result.

His final position in the Canadian Government was as Deputy Project Manager, head of the On-site team, for the construction of a series of vessels for a government agency which operates ships. The on-site team compiled a large list of issues. Many were minor, but some were serious, such as: non-compliance with the collision regulations, high voltage electrical equipment in machinery spaces not in compliance with approved plans, a lack of structural fire protection in the aluminium deckhouse under the bridge (ships built to the same design in about a dozen other countries had a 30 minute fire rating, Canada had none), bulkhead glands not packed properly, elimination of roll reduction devices, and so on. He and the On-site team were ordered not to report safety concerns to headquarters or to the fleet. Once he also assumed the role of union local president, he raised these issues formally under the *Canada Labour Code*. With the active support of the national union president, he met with senior government executives in Ottawa. Some concerns were addressed, but most, including the lack of structural fire protection, were not. Once again, his employment was terminated – at some cost to the Canadian taxpayer.

Figure 6: Corrosion in a Canadian vessel
"Class certified and Transport Canada approved"

Dalziel was a sophisticated, knowledgeable professional. Despite this, he was unable succeed in a country with a reputation for government excellence, low levels of corruption, and strong rule of law. This suggests that risks to ordinary sailors and other personnel may be unsurmountable.

3.7 THE COMMON THREAD

The safety culture issues are not restricted to the maritime industries. They are a much wider societal problem.

3.7 (a) Rail Safety

In the evening of July 6, 2013, an unmanned train, loaded with 6,000 tons of mis-declared oil in substandard rail tank cars with poorly maintained locomotives rolled down a long steep hill into the town of Lac Mégantic, Quebec, derailed, exploded, eviscerated the town and incinerated 47 people.

Figure 7: Book cover [36]

Why did this happen? The former area superintendent of the railroad had objected to the operational changes being made, he was demoted and sidelined, and eventually left. Bruce Campbell details that the tragedy was the consequence of four decades of mutually reinforcing policies – deregulation, privatization, tax cuts and austerity – that eroded transportation safety protections. The entrenched power of railway corporations acted in concert with government enablers and regulators, a phenomenon known as "regulatory capture" [36], [37].

Also of great concern, the chapter 'The Investigation that Lost Heart' reveals how the Canadian Transportation Safety Board (TSB) report went through several iterations, removing blame that may be associated to government regulators. However, in the press conference releasing the report, Wendy Tadros, Chair of the TSB, gave a withering condemnation of Transport Canada's failures; her final act as Chair of the Transportation Safety Board [36].

3.7 (b) Mine Safety

On May 9, 1992, the Westray Coal Mine in Nova Scotia exploded. In 2004, after continued Union pressure, the *Westray Act* was passed, to hold management accountable, in the 20 years since it has seldom been effectively enforced.

"Today, we remember the 26 miners who perished 32 years ago because of what a judge described as 'a complex mosaic of actions, omissions, mistakes, incompetence, apathy, cynicism, stupidity and neglect,'" said Marty Warren, Canadian National Director of the United Steelworkers Union. He added "Since the enactment of the *Westray Act*, over 18,000 workers have lost their lives due to workplace hazards. Many thousands more have had their lives forever altered by workplace injury and occupational illness... This staggering number highlights the urgent need for robust enforcement of the law. Every worker deserves to return home safely to their families at the end of their workday" [38].

3.7 (c) Food Safety

Dr. Yasmine Motarjemi is a former senior scientist at World Health Organization. Head hunted by the Management of Nestlé in the year 2000, she was appointed as Assistant Vice President, in charge of food safety globally. Shortly after joining the company, her administrative assistant gave her a sculpture of the three wise monkeys symbolizing "see no evil, hear no evil, speak no evil" [39].

Figure 8. The Three Wise Monkeys presented to Dr. Motarjemi [40]

Eventually, after four years of internal whistleblowing, she spent over 12 years before the Swiss Courts, trying to prove the retaliation she experienced and defend the right to maintain food safety standards and to protect public health. A colleague wrote her: "We all know, but we keep quiet because we have families to feed and pensions to secure" [39].

Dr. Motarjemi noted, "At the end of the story... whistleblowing remains a lonely journey" [41].

3.7 (d) Air Safety

The ongoing safety issues at Boeing are well publicized. Boeing is one of the largest and most important companies in the United States. Arguably, it is too big to fail. But is it also too big to be held to account? Senator Richard Blumenthal, a Democrat who has led hearings focused on Boeing's retaliation against whistleblowers said, individuals, not just the company, should face consequences "for past illegalities as well as continuing retaliation against whistleblowers and other wrongdoing" [42]. In the meantime, two Boeing whistleblowers have died, one allegedly by suicide [43].

As Dr. Motarjemi and the Boeing whistleblowers have made clear, a whistleblower case is very personal. It is also very personal for those who suffer because of the incidents that whistleblowers tried to prevent.

Figure 9: Some Boeing Max 8 victims, Ethiopia, March 2019 [44]

4. WHAT MAKES A ROBUST SPEAK-UP / WHISTLEBLOWING SYSTEM

The logic of whistleblower protections is not complex: First, develop and implement a policy legitimizing the disclosure of wrongdoing outside the chain of command. Second, conduct investigations into disclosures of wrongdoing and act constructively on the findings, while at the same time protecting the whistleblowers. This will create trust in the regime, which will increase use, facilitate the quick correction of wrongdoing, and deter future wrongdoing.

4.1 COMPONENTS OF EFFECTIVE DISCLOSURE REGIMES

Organizations sincerely wishing to develop a whistleblowing system are no longer pioneers but can rely on existing best practices. Moreover, to be effective, a speak-up system should be one element of an ethics, integrity, or compliance regime. Using Kaptein's [19] model, this should include a code of ethics accompanied by training and consistent communication, accountability policies, monitoring and auditing, investigation and correction policies, a dedicated ethics officer providing advice, an ethics report line, and policies to incentivize ethical conduct. Whistleblowing procedures are consistent with the reporting, investigation, and correction aspects of this model. Robust whistleblowing systems have hard (structural) and soft (cultural) components.

Although this paper cannot cover all the hard components of an effective disclosure system, standards and guides have been developed by non-governmental organizations. Prominent examples include the U.S. Government Accountability Project (GAP) *International Best Practices* [45], Transparency International's [46] *Assessing Whistleblowing Legislation, ISO 37002: Whistleblowing Management Systems* [20], the Blueprint for Free Speech *Getting Whistleblower Protection Right* [47], and the Centre for Free Expression *Evaluation Criteria for Protection of Whistleblowers* [48]. The European Union Directive [49] on whistleblowing incorporates most best practices. Also helpful is *The Whistleblowing Guide: Speak-up Arrangements, Challenges and Best Practices* [50]. Despite the differences in these best practice documents, all address certain key principles:

- A broad scope of protection, ensuring that any worker (even contract personnel) can raise a concern on any issue without interference or penalty. There should be multiple reporting avenues and very few restrictions on how a disclosure is made.
- Protections that start as soon as the whistle is blown, or even before (such as when seeking advice). Whistleblowing may begin as a casual observation; wrongdoers will not wait for a formal report to begin reprisals.
- Reliable and prompt avenues for redress if reprisal occurs. This benefits the organization by returning the whistleblower to a productive role quickly. It should cover all costs the whistleblower has been forced to assume and should not extend the process for years.
- Robust mechanisms for investigating and correcting misconduct. Whistleblowers are deterred from speaking up if they believe it will not lead to any improvement. Important aspects include impartial, competent investigations, prompt correction of the misconduct, and consequences for wrongdoers.
- A cycle of monitoring and improvement. This includes identifying performance indicators and collecting data that accurately reflect the effectiveness of the policy.

Considering the soft components of a reporting mechanism, leadership commitment to open communication is an important factor [51]. This is because leaders establish and provide (or withhold) resources for any reporting mechanism, as well as training in its use. When perceptions of organizational culture and organizational justice are positive, internal whistleblowing is more common. Weaker ethical cultures, on the other hand, are associated with external whistleblowing and inaction [52]–[54].

Organizational leaders should take note: Observers of wrongdoing take cues from their environment to determine whether there is an open culture in which they can speak truth to power. If it is not safe to speak up internally, they may choose to remain silent or to blow the whistle externally [53], [54]. Additionally, external whistleblowing is more likely if the individual has already suffered reprisals or is angry at inaction by management. In effect, organizational leaders can push employees into becoming whistleblowers, creating reputational damage they would presumably wish to avoid.

4.2 COMMON FLAWS IN WHISTLEBLOWING REGIMES

There are several risks that can undermine even the best policies. The first is perhaps the most obvious: "…If the weakest link… is an unwilling manager, corrupt CEO or incompetent board, then any kind of low courage whistleblowing channel will remain ineffective" [55, p. 317]. Even where leadership is not directly implicated, however, leadership can neglect the regime or apply it inconsistently. It is common for leaders to be sceptical of disclosures and to focus on those making the disclosure. This, in turn, can lead to ignoring disclosures or minimizing the potential seriousness of the issue disclosed. In addition, organizations can place too much emphasis on correcting the wrongdoing and not enough on protecting the whistleblower. This may reflect the design of policies and procedures but may also reflect bias. However, it undermines trust in the regime. Other ways to erode trust include excessive loyalty to organizational reputation and secretiveness [56].

Operational pressures can also be a significant threat. This was found in governments, where senior officials may regard whistleblowing mechanisms as a nuisance and burden [57, p. 145]. It may be even truer in the private sector, where pressures to produce profit create a temptation to make trade-offs with safety:

…Not all companies are managed well or staffed with the resources necessary to support the master in maintaining a safely operated ship. In these cases, the reporting of deficiencies may be looked upon as a problem rather than as an opportunity to improve safety. In such companies, a master who brings safety management problems to the company is himself a problem and risks being replaced. This can have a chilling effect on other masters who then become reluctant to bring their own safety concerns to management. [34, p. 3]

John Loftus, who was Master of the *Horizon Trader*, was fired after he reported safety concerns. He sued and was awarded more than $1.15 million in back wages, damages, and costs [58]. Pressure may also be applied to government regulators by industry associations; this is exacerbated by the fact that the IMO is not a regulatory body, and flag states compete with one another. Flag states may then delegate most of their responsibilities to ROs such as classification societies, creating two levels of conflict of interest:

The RO class societies are employed by companies but act on behalf of flag states in implementing and enforcing international regulations. Companies, in turn, employ RO class societies to prepare their ISM Code safety management systems and conduct audits of SMS documentation and performance. The end result is that companies are regulated by private organizations that they themselves employ and have a choice in selecting. [34, p. 3]

The fact that this works well in most instances should not invite complacency; indeed, it underlines the importance of a robust SMS with safe, effective reporting avenues.

5. RECOMMENDATIONS

In Canada and around the world, safety cultures remain undeveloped: time after time, profits, schedules, personal careers, and political priorities are put ahead of safety. There are three key areas where the need for change is urgent:

- Prevention of regulatory capture,
- Whistleblower protection, and
- Legal accountability for senior management and government.

Each disastrous regulatory failure is unique, and yet everyone could have been prevented if warnings were heeded and the regulations already in place were enforced. For all the diversity of their technical causes, the root cause of all these disasters is the same: corporate negligence facilitated by regulatory failure, and worse, regulatory capture. Considering evidence that the ISM Code is not achieving its goals, and to help preserve life, the environment, and the public interest, the authors recommend that:

- The IMO revise the ISM Code to include an effective whistleblower protection regime, and that national legislation be updated in accordance with the ISM Code. Stakeholders must include unions and professional societies to ensure that resulting regimes reflect reality faced by workers.
- Port State Control regimes look for evidence of an effective whistleblower protection regime, and when one is not found, take appropriate action.
- The International Transport Workers Federation and professional societies provide guidance to their members and, if necessary, legal and financial support (they should endeavour to recover the cost of such support from the companies involved).
- The International Association of Classification Societies require its members have an effective whistleblower protection regime
- Insurers look for an effective whistleblower protection regime when establishing insurance rates.
- Governments establish and enforce criminal statutes, backed by effective enforcement regimes, that expand accountability from companies to the individual executives, managers and supervisors unless they can demonstrate that they had no prior knowledge of the factors leading to the incident and have an effective whistleblower protection regime.
- Civil courts allow financial claims to be heard against the responsible executives, managers and Supervisors.

6. CONCLUSION

The Regulatory Regime Exists
The ISM Code requires safety issues be reported.
However, report them and you may be fired.
IT IS A REGULATORY CATCH-22. [34]

The loss of the *Herald of Free Enterprise* off Zeebrugge prompted the development of the ISM Code, which should have led to improvements in safety. However, as the cases in this paper demonstrate, disasters have continued in the marine environment, as well as in other transportation sectors. Common links include commercial pressures, regulatory capture, and concerned employees either failing to speak up or being ignored. This is not an accident: organizational leaders create the conditions in which personnel do not trust that concerns will be addressed and believe they will be punished if they speak up.

This is counterproductive: experience has shown the cost of failing to implement robust whistleblower protections: problems fester, unreported, until the piper must eventually be paid – the bill may include reputational damage, millions in repairs, billions in environmental damage and lives lost. Flag and port states must also recognize that when that failure comes, it may be in their waters or ports, and it may be their citizens [59]. Accordingly, it is in their interests to ensure that they use the best resource at their disposal: loyal employees and other workers, who know where hazards lie and want them corrected – not least because their lives may depend on it.

"The best set of eyes and ears to ensure safety and regulatory compliance for any ship is its crew." [34]

"Every lie we tell incurs a debt to the truth. Sooner or later that debt is paid." ('Chernobyl' miniseries, 2019)

7. REFERENCES

1. United Kingdom. Department for Transport, "MV Herald of Free Enterprise: Formal investigation," HM Stationery Office, London, 8074, 1988.

2. NEAR, J P., and MICELI, M P., "Organizational dissidence: The case of whistle-blowing," *J. Bus. Ethics*, vol. 4, no. 1, pp. 1–16, 1985.

3. MILLIKEN, F J., MORRISON, E W., and HEWLIN, P F., "An exploratory study of employee silence: Issues that employees don't communicate upward and why," *J. Manag. Stud.*, vol. 40, no. 6, pp. 1453–1476, 2003.

4. VANDEKERCKHOVE, W., and COMMERS, M. S. R., "Whistle blowing and rational loyalty," *J. Bus. Ethics*, vol. 53, no. 1–2, pp. 225–233, 2004.

5. LACHMAN, V D., "Whistleblowing: Role of Organizational Culture in Prevention and Management," *Ethics, Law, Policy*, vol. 17, no. 4, pp. 265–267, 2008.

6. DONKIN, M., SMITH, R., and BROWN, A J., "How do officials report? Internal and external whistleblowing," in Whistleblowing in the Australian public sector: Enhancing the theory and practice of internal witness management in public sector organisations, A. J. Brown, Ed. Canberra: ANU E-Press, 2008, pp. 83–106.

7. MESMER-MAGNUS, J R., and VISWESVARAN, C., "Whistleblowing in organizations: An examination of correlates of whistleblowing intentions, actions, and retaliation," *J. Bus. Ethics*, vol. 62, no. 3, pp. 277–297, 2005.

8. MICELI, M P., NEAR, J P., and DWORKIN, T M., *Whistle-blowing in organizations*. New York: Routledge, 2008.

9. LEWIS, D., "Ten years of Public Interest Disclosure Act 1998 claims: What can we learn from the statistics and recent research?," *Industrial Law Journal*, vol. 39, no. 3. pp. 325–328, 2010.

10. REASON, J., "Achieving a safe culture: Theory and practice," *Work Stress*, vol. 12, no. 3, pp. 293–306, 1998.

11. URBINA, I., "Workers on Doomed Rig Voiced Safety Concerns," *New York Times*, vol. 2, New York, pp. 2–5, 21-Jul-2010.

12. Association of Certified Fraud Examiners, "Occupational fraud 2022: A report to the nations," 2022. Available: https://acfepublic.s3.us-west-2.amazonaws.com/2022+Report+to+the+Nations.pdf

13. FELDMAN, Y., and LOBEL, O., "Decentralized enforcement in organizations: An experimental approach," *Regul. Gov.*, vol. 2, no. 2, pp. 165–192, 2008.

14. CALLAHAN, E S., DWORKIN, T M., FORT, T L., and SCHIPANI, C A., "Integrating trends in whistleblowing and corporate governance: Promoting organizational effectiveness, societal responsibility, and employee empowerment," *Am. Bus. Law J.*, vol. 40, no. 1, pp. 177–215, 2002.

15. DWORKIN, T M., and BAUCUS, M S., "Internal vs. external whistleblowers: A comparison of whistleblowing processes," *J. Bus. Ethics*, vol. 17, no. 12, pp. 1281–1298, 1998.

16. ALLINGHAM, J., "B.C. health researcher firing scandal timeline," *CBC News*, 13-Jun-2015. Available: https://www.cbc.ca/news/canada/british-columbia/b-c-health-researcher-firing-scandal-timeline-1.3109898.

17. JOHANNESEN, N., and STOLPER, T., "The Deterrence Effect of Whistleblowing: An Event Study of Leaked Customer Information from Banks in Tax Havens," CESifo Group, Munich, 6784, 2017.

18. VAN DYNE, L., and LEPINE, J A., "Helping and voice extra-role behaviors: Evidence of construct and predictive validity," *Acad. Manag. J.*, vol. 41, no. 1, pp. 108–119, 1998.

19. KAPTEIN, M., "The effectiveness of ethics programs: The role of scope, composition, and sequence," *J. Bus. ethics*, vol. 132, no. 2, pp. 415–431, 2015.

20. ISO, "ISO 37002: Whistleblowing management systems," 2021.

21. VANDEKERCKHOVE, W., KENNY, K., FOTAKI, M., and HUMANTITO, I., "Effective speak-up arrangements for whistle-blowers: A multi-case study on the role of responsiveness, trust and culture," London, 2016.

22. BROWN, A J., and WHEELER, C., "Project findings: An agenda for action," in Whistleblowing in the Australian public sector: Enhancing the theory and practice of internal witness management in public sector organisations, A. J. Brown, Ed. Canberra: *ANU E-Press*, 2008, pp. 289–312.

23. ANDREW, W., JAMES, R., and HANDFIELD, M., "Royal Commission of Inquiry into the sinking of the MV Princess Ashika," Nuku'Alofa, Tonga, 2010. Available: https://natlib.govt.nz/records/21644069

24. Transport Accident Investigation Commission, "Final Preliminary Report Prepared by the Transport Accident Investigation Commission for The Royal Commission of Inquiry into the Sinking of the MV Princess Ashika," Nuku'alofa, Jul. 2010. Available: https://www.taic.org.nz/sites/default/files/inquiry/documents/09-209%20Final%20Preliminary%20Report.pdf

25. "Tonga witness feared dismissal over Ashika concerns," *RNZ News*, 17-Feb-2011. Available: https://www.rnz.co.nz/international/pacific-news/195426/tonga-witness-feared-dismissal-over-ashika-concerns..

26. CASS, P., "'Any fool could tell' – but still the ferry fit only for scrap was allowed to sail to her doom," *Kaniva Tonga News*, 31-Jul-2022.

27. CHAPIN, L., "Corrosion missed during inspections grounded P.E.I. ferry, documents show," *CBC News*, 09-Jun-2022. Available: https://www.cbc.ca/news/canada/prince-edward-island/pei-ferry-holiday-island-ballast-steel-inspection-transport-canada-1.6472050.

28. GALLANT, I., "MV Holiday Island to be scrapped due to major fire damage," *CBC News*, 14-Oct-2022. Available: https://www.cbc.ca/news/canada/prince-edward-island/pei-holiday-island-ferry-disposal-1.6616352.

29. DODD, S., CAMPBELL, B., and DALZIEL, J., "The Ocean Ranger disaster taught us much but lack of action by governments, corporations still fails workers," *Saltwire*, 14-Feb-2022. Available: https://www.saltwire.com/atlantic-canada/opinion/commentary-the-ocean-ranger-disaster-taught-us-much-but-lack-of-action-by-governments-corporations-still-fails-workers-100693052/.

30. FRUMP, R., Until the sea shall free them: Life, death, and survival in the Merchant Marine. New York: *Doubleday*, 2002.

31. FRUMP, R., The Captains of Thor: What really caused the loss of the SS El Faro in Hurricane Joaquin. *Race Point*, 2018.

32. PELLEY, S., "'Incredibly moving' audio from El Faro's final hours," *CBS News*, 05-Mar-2017. Available: https://www.cbsnews.com/news/incredibly-moving-audio-from-el-faros-final-hours/. [Accessed: 03-Aug-2024].

33. US. Department of Homeland security, "Steam Ship El Faro (O.N. 561732) sinking and loss of the vessel with 33 persons missing and presumed deceased northeast of Acklins and Crooked Island, Bahamas on October 1, 2015," 2017.

34. LOFTUS J., *et al.*, "Spotlight on safety: Why accidents are often not accidental," 2019. https://bridgedeck.org/actionnetwork/Spotlight-MMP-WMRC-Final071619.pdf

35. DALZIEL, J., "Barracudina: Lost but not forgotten," *The Maritime Executive*, 25-Nov-2018. Available: https://www.maritime-executive.com/editorials/barracudina-lost-but-not-forgotten.

36. CAMPBELL, B., The Lac-Megantic rail disaster: Public betrayal, justice denied. Toronto: *James Lorimer & Company Ltd*, 2018.

37. CAMPBELL, B., "Disasters foretold: Boeing 737 Max 8 and Lac-Mégantic," *The Conversation*, 04-Feb-2020.

38. United Steelworkers, "No more Westrays," 09-May-2024. Available: https://usw.ca/westray-act-failing-slain-workers-20-years-after-adoption/.

39. TURNER, J., "Dr. Yasmine Motarjemi," *Whistleblower Network News*, 25-May-2022. Available: https://whistleblowersblog.org/whistleblower-of-the-week/dr-yasmine-motarjemi/.

40. MOTARJEMI, Y., "Challenges of Whistleblowers - Episode 4: 'Silence- the sound of violence,'" *LinkedIn*, 05-Nov-2023. Available: https://www.linkedin.com/pulse/challenges-whistleblowers-episode-4-silence-sound-yasmine-motarjemi-zpyee/.

41. MOTARJEMI, Y., "Whistleblowing: Time for international Switzerland to meet the global standard," *Global Geneva Insights*, 17-Aug-2020. Available: https://global-geneva.com/whistleblowing-time-for-international-switzerland-to-meet-the-global-standard/.

42. SHERMAN N., and LEGGETT, T., "Boeing: What's the right punishment?," *BBC*, 08-Jul-2024. Available: https://www.bbc.com/news/articles/c6p2jeg14r9o.

43. CIMINO, A., "Boeing whistleblower's autopsy revealed: John Barnett was found dead in his truck with gun in hand next to a suicide note," *Daily Mail Online*, 17-May-2024. Available: https://www.dailymail.co.uk/news/article-13431401/boeing-whistleblower-autopsy-john-barnett-dead-suicide.html.

44. VERSPRILLE, A., "Boeing 737 Max crash victims' families object to plea deal," *BNN Bloomberg*, 31-Jul-2021. Available: https://www.bnnbloomberg.ca/business/2024/07/31/boeing-737-max-crash-victims-families-object-to-doj-plea-deal/.

45. DEVINE, T., "International best practices for whistleblower policies," *Government Accountability Project*, 2016. Available: https://whistleblower.org/international-best-practices-for-whistleblower-policies/.

46. MCDEVITT A., and TERRACOL, M., "Assessing whistleblowing legislation," Transparency International, Berlin, 2020.

47. COLVIN, N., GALIZZI, B., and NAD, V., "Getting whistleblower protection right: A practical guide to transposing the EU Directive," 2020.

48. BRON, I., and HUTTON, D., "Evaluation criteria for protection of whistleblowers: A guide for legislation and policy," Ottawa, 2022.

49. EUROPEAN UNION, "Directive (EU) 2019/1937 of the European Parliament and of the Council of 23 October 2019 on the protection of persons who report breaches of Union law." Brussels, 2019.

50. KENNY, K., VANDEKERCKHOVE, W., and FOTAKI, M., *The whistleblowing guide: Speak-up arrangements, challenges and best practices*. Newark: John Wiley & Sons, 2019.

51. DETERT., J R and BURRIS, E R., "Leadership behavior and employee voice: Is the door really open?," *Acad. Manag. J.*, vol. 50, no. 4, pp. 869–884, 2007.

52. MICELI, M P., and NEAR, J P., "Characteristics of organizational climate and perceived wrongdoing associated with whistle-blowing decisions," *Pers. Psychol.*, vol. 38, no. 3, pp. 525–544, 1985.

53. ROTHSCHILD, J., and MIETHE, T D., "Whistle-blower disclosures and management retaliation: The battle to control information about organization corruption," *Work Occup.*, vol. 26, no. 1, pp. 107–128, 1999.

54. KAPTEIN, K., "From inaction to external whistleblowing: The influence of the ethical culture of organizations on employee responses to observed wrongdoing," *J. Bus. Ethics*, vol. 98, no. 3, pp. 513–530, 2011.

55. VANDEKERCKHOVE, W., BROWN, A J., and TSAHURIDU, E., "Managerial responsiveness to whistleblowing: Expanding the research horizon," in *International handbook on whistleblowing research*, A. J. Brown, D. Lewis, R. Moberly, and W. Vandekerckhove, Eds. Northampton, Mass.: Edward Elgar, 2014, pp. 298–327.

56. ROBERTS, P., OLSEN, J., and BROWN, A J., "Whistling While They Work: A good-practice guide for managing internal reporting of wrongdoing in public sector organisations," Canberra, 2011.

57. BRON, I., "Square peg in a round hole? Three case studies into institutional factors affecting public service whistleblowing regimes in the United Kingdom, Canada, and Australia," Carleton University, 2022.

58. GOETSCH, C., "Whistleblowing: Ship master wins large settlement in US court," *gCaptain*, 19-Jul-2016. Available: https://gcaptain.com/whistleblowing-ship-master-wins-large-settlement-in-us-court/.

59. DALZIEL, J., "Remembering the Leros Strength," *The Maritime Executive*, 25-Feb-2017. Available: https://maritime-executive.com/editorials/remembering-the-leros-strength

8. AUTHORS' BIOGRAPHY

Ian Bron, M.Ed., PhD., is a Senior Fellow at the Toronto Metropolitan University Centre for Free Expression with experience as a government regulatory specialist, evaluator, educator, and naval officer. His interest in whistleblowing originated from personal experience after he raised concerns about the oversight of marine transportation security by Canada's Ministry of Transport in 2006. Reprisals drove him from the public service. Since then, he has been active in advocacy and has provided analysis and testimony to legislative committees. He is interested in the design and evaluation of systems for whistleblowing, ethics, and accountability in organizations and society.

Ronald Pelot, PhD., P.Eng., has been a Professor in the Department of Industrial Engineering at Dalhousie University. In 1997, he founded the Maritime Activity and Risk Investigation Network at Dalhousie. His team has developed new software tools and analysis methods applied to maritime safety (accidents), coastal zone security, and environmental impacts of shipping. Research methods encompass spatial risk analysis, vessel traffic modelling, data processing and pattern analysis, location models for resource allocation, and safety analyses. Recent projects include developing models to refine the methods for calculating the Maximum Expected Time of Rescue in the POLAR Code. Studies are underway to improve SAR response capabilities and effectiveness in the Canadian North. Dr. Pelot has published over 50 journal articles and produced more than 100 technical reports.

John W. Dalziel, M.Sc., P.Eng., MRINA, is an Adjunct Professor in the Department of Industrial Engineering, Dalhousie University, Halifax, Canada. He has also taught naval architecture and management accounting at the technical college/university level. He studied Naval Architecture in the United States and Scotland, and management science in England. He has sailed through the Arctic with the Canadian Coast Guard, and with the fishing industry. Over the past half century, most of his experience has been in the supervision of ship construction, refits, and repairs, and in safety inspection and survey on behalf of flag state and class. He has had three whistleblowing cases, one reported internationally and raised in Parliament. He has spoken internationally on maritime safety and search and rescue.

Human Factors 2024
8th – 9th October 2024
Wageningen, Netherlands

HOW CAN WE SUPPORT UNCREWED MARITIME AUTONOMOUS SURFACE SHIP OPERATORS IN THEIR DECISION-MAKING?

K M Lynch*, **M S Young, D Taunton,** University of Southampton, UK, **A P J Roberts, V A Banks,** Thales, UK and **K L Plant,** University of Southampton, UK

SUMMARY

The operation of uncrewed Maritime Autonomous Surface Ships (MASS) may make decision-making for operators more difficult due to their relocation to a Remote Control Centre (RCC), therefore they will lack proximity to the MASS and may have reduced situational awareness. The Schema World Action Research Method (SWARM) has been used to explore the decision-making process of operators and the associated challenges with their operation. Seven operators and four subject matter experts were interviewed using SWARM about their decision-making. The interview responses were coded using the Schema Action World (SAW) Taxonomy and the results were then grouped into themes. Design principles for decision-making were then generated from the SAW coding groups. The design principles included design considerations such as the system's transparency, the types of display indications needed and designs to support system monitoring, and training and experience for the development of appropriate schema.

NOMENCLATURE

MASS	Maritime Autonomous Surface Ship
PCM	Perceptual Cycle Model
RCC	Remote Control Centre
SAW	Schema Action World
SWARM	Schema World Action Research Method

1. INTRODUCTION

Maritime Autonomous Surface Ship (MASS) operators will have challenges in making decisions when operating uncrewed MASS from Remote Control Centres (RCC; degree three) due to their lack of proximity and working with highly automated systems [1-3]. MASS has various applications such as hydrographic and bathymetric surveys [4, 5], search and rescue [6] and intelligence and surveillance [7]. At the RCCs, operators will be predominantly 'on the loop' where their main actions will be supervising the automated systems and taking back control if the automated system has an issue or cannot handle a situation [8].

One of the challenges for operators' decision-making will be maintaining their situational awareness as they will lack the physical cues they would have when operating onboard a conventional vessel [2, 9, 10]. Situational awareness is important for decision-making to ensure that the decision-maker has an accurate understanding of the situation, otherwise it could lead to inappropriate decisions [11, 12]. A complication for uncrewed MASS, is that operators will be reliant on the information provided by their technological systems to inform their situational awareness [3]. Operators will utilise information for similar systems as a conventional vessel to navigate the MASS such as, an Electronic Chart Display Information System (ECDIS) and radar, the difficulty however is they will not have the physical cues to verify the digital information being presented to them [2]. As operators will be using solely digital information it will important that they are provided with necessary level of information to inform their decisions without causing cognitive overload [13].

Another difficulty will be operators maintaining their situational awareness and remaining vigilant to changes whilst monitoring the automated systems, due to the issues associated with operators continuously monitoring automated systems for long periods [14-16]. Especially in the case where operators are expected to be passive monitors of the automated system and only take control or intervene when the MASS' automated system has a problem [17]. Passive monitoring behaviour could cause a decision latency as operators will take time to get back into the loop before they can make an informed decision [2, 18]. Delays caused by communications links could cause latency in the information being received from the MASS and could also affect the operator's decision-making time, which will be important when responding to takeover requests or situations [19, 20].

This paper aims to investigate the key factors involved in MASS operators' decision-making at RCCs and how their decision-making processes can be supported. The Schema World Action Research Method (SWARM) developed by Plant and Stanton (2016) will be used as an interview schedule to interview MASS operators and Subject Matter Experts (SMEs), to explore the information and experiences they utilise when making decisions. The results of the decision modelling will inform design principles to support operator decision-making at RCCs and identify further challenges for decision-making.

1.1 THE PERCEPTUAL CYCLE MODEL

Naturalistic decision-making is characterised by uncertainty, time pressures, high risks, competing goals and changeable conditions [21]. Figure 1 shows the naturalistic decision-making model the Perceptual Cycle Model (PCM; Neisser, 1976). The PCM is a framework that can be used to show an operator's decision-making process, it links together three categories: schema, action and world. Schema refers to the operator's cognitive map of the world, which is formed from their past experiences and the world information. World is the information that is available to the operator, this could be information from their user interface or information communicated to them from a team member or the natural environmental conditions. Actions, the actions the operator takes e.g., taking back manual control, changing the vessel's path or monitoring the vessel's communication link. The PCM shows how the information in the world modifies the operator's schema (their mental model of the situation) and then how their schema directs the actions they then take in which they will sample further world information.

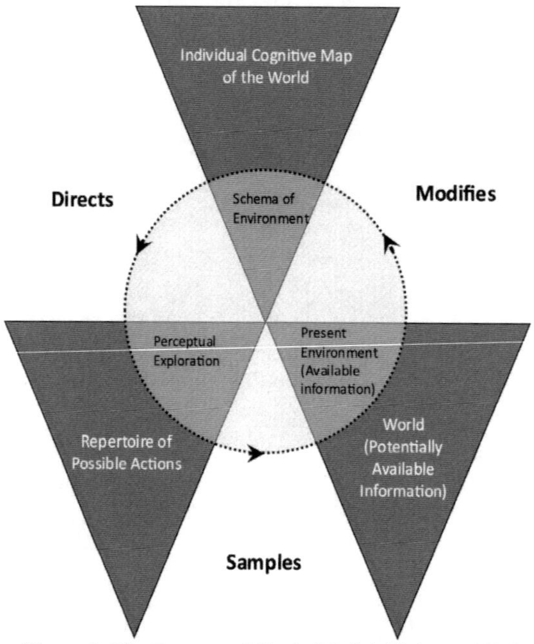

Figure 1: The Perceptual Cycle Model (Neisser, 1976).

2. METHOD

2.1 Schema World Action Research Method

SWARM was developed to capture how pilots make critical decisions and gain insights into their decision-making processes by applying the PCM framework to their decision-making [22]. SWARM is an interview method that uses cognitive prompts for each of three PCM components (Schema, Action and World) to investigate how decisions are made. SWARM captures data on what information the decision-maker is using in their environment to shape their understanding of a situation. It also shows how they use their experiences and previous knowledge to take action and make decisions. SWARM has previously been used to generate design requirements for future avionic systems to better support pilot decision-making in the event of an engine failure during take-off or a suspected engine oil leak [23, 24].

Although SWARM was developed to analyse aeronautical decision-making the PCM framework has been used to explore decision-making in other safety-critical domains including uncrewed vehicle operations and other automated systems where the operators are operating a vehicle remotely or the operator's role is mainly supervisory [25-27]. The application of SWARM to MASS operator decision-making will be used to explore the interaction between the operator and automated system and find the key factors in decision-making when operating MASS, including how they respond to potentially critical incidents and make decisions with the automated systems.

2.2 Participants

Eleven participants took part in the SWARM interviews and Table 1 shows the key demographics of the participants. Due to the infancy of MASS and therefore a limited number of operators, the participant recruitment criteria also included participants without direct operational experience, but they knew about the development of current and future systems. Four of the participants had knowledge of MASS development, trials and certification and the other seven participants had direct operational experience.

Table 1. Key participant demographics. (*one participant had a significantly higher number of operational hours so was excluded from the average number of operational hours)

Gender	Male	9
	Female	2
Age (Years)	Average	41.82
	Range	21-68
Years of Maritime Experience	Average	18.68
	Range	2.5-40
Operational Experience (Years)	MASS Operator	7
	Subject Matter Expert	4
Operational Hours	Average (n=6)	288
	Range (n=6)	180-600
	Operational Hours* (n=1)	50,000

2.3 SWARM Interviews

The interviews were conducted in person or on video calls using Microsoft Teams, which was used to record and generate transcripts of all the interviews. Transcripts were checked, amended as necessary and anonymised by the researcher. The completed transcripts were then qualitatively analysed using the data analysis software Nvivo 12.

Before starting the interview, participants were given background to decision-making for MASS operations and the Human Factors challenges surrounding their remote operation and an overview of the study procedure. Next, the interviewer gave the participants an explanation of the PCM and how it can be applied to investigate an operator's decision-making process. Then the SWARM interview was conducted, and the participants were asked the SWARM prompts in Table 2 and additional follow-up questions were needed to gain an insight into how they made their decisions.

Table 2. SWARM prompts selected for the interviews [22].

SAW Theme	SWARM Prompt
Schema	1. Does this use case remind you of anything you've read or heard or been involved in? 2. Would you draw on previously acquired knowledge of similar MASS or other uncrewed vehicle operations? 3. Could you draw on previous experiences of similar situations of conducting a survey? (if yes, what) 4. Would your previous direct experiences influence you? (if yes, how) 5. What pre-existing training that you've had might be useful in this case? 6. What information would you know to be true or were you certain of? (Why?) 7. Would you create a mental image in your mind to assist you? • (If yes) What information (mental and/or physical) would you utilise to do this? 8. Would you imagine the consequences of your actions? • (If yes) Would this impact on or change the decision/actions that you made?
Action	1. Thinking about the use case – what would the main actions be? 2. What information (mental and/or physical) would you be most useful or you would use to assist with your actions? 3. What navigational actions might you take? 4. What information (mental and/or physical) would you use to help you navigate? 5. Would you communicate with anyone? (If yes, who to / what would be communicated?) 6. In relation to the technological system, what would you be looking at (observing or checking) during the use case? What information would be essential to include?

	7. In relation to the physical environment, what might you look at (observe or check)? 8. What were the key decisions might you make?
World	1. What artefacts would you have available to you? (e.g. physical objects, equipment, written documents, etc.) 2. What information from the technological system(s) would you utilise/be most useful? 3. How and where would you get this information? 4. Might you receive information from others and how would you receive this information? (If yes, what) 5. What could the communicated information tell you?

2.4 Data Analysis

The SAW taxonomy, shown in Table 3 developed by Plant and Stanton [28] was then used to code the transcripts to explore the relevant factors involved in MASS operators' decision-making for each of the components of the PCM. The taxonomy was adapted for MASS operations to remove references to aviation to reference MASS instead and one code, 'Non-natural environmental conditions' was added to describe other environmental conditions such as, other vessels, navigational lights and markers. The SAW coding results were then grouped into themes within each code and then design principles were generated for each of the 27 codes found in the transcripts.

Table 3. SAW Taxonomy adapted for MASS operations [28]

Schema Themes	**Action Themes**	**World Themes**
Analogical Schema	Operate	Absent information
Declarative Schema	Communicate	MASS status
Direct Past Experience	Decision action	Artefacts
Insufficient Schema	Environmental monitoring	Communicated information
Trained Past Experience	Navigate	Display indications
Vicarious Past Experience	Non-action	Location
	Situation Assessment	Natural environmental conditions
	Standard operating procedure	Non-natural environmental conditions
	System Interaction	Operational context
	System Monitoring	Physical cues
		Severity of problem
		Technological conditions

3. RESULTS

3.1 SAW Coding

Figure 2 shows the results of the SAW coding for each sub-category. The SAW category with the highest number of references was World (53% of the total references), followed by Action (32%) and then Schema (15%). The top overall six SAW codes with the highest number of references were 'Display Indications' (n=105), 'Artefacts' (n=86), 'System Monitoring' (n=74), 'Natural Environmental Conditions' (n=56), 'Navigate' (n=55) and 'Operational Context' (n=55). The Schema codes with the highest number of references were the 'Declarative Schema' and 'Direct Past Experience' codes (21% and 17% of the total schema references respectively). In the Action category, 'System Monitoring' and 'Navigate' codes had the highest number of references (24% and 18%). The World codes with the highest references were the 'Display Indications' and 'Artefacts' codes (34% and 30%).

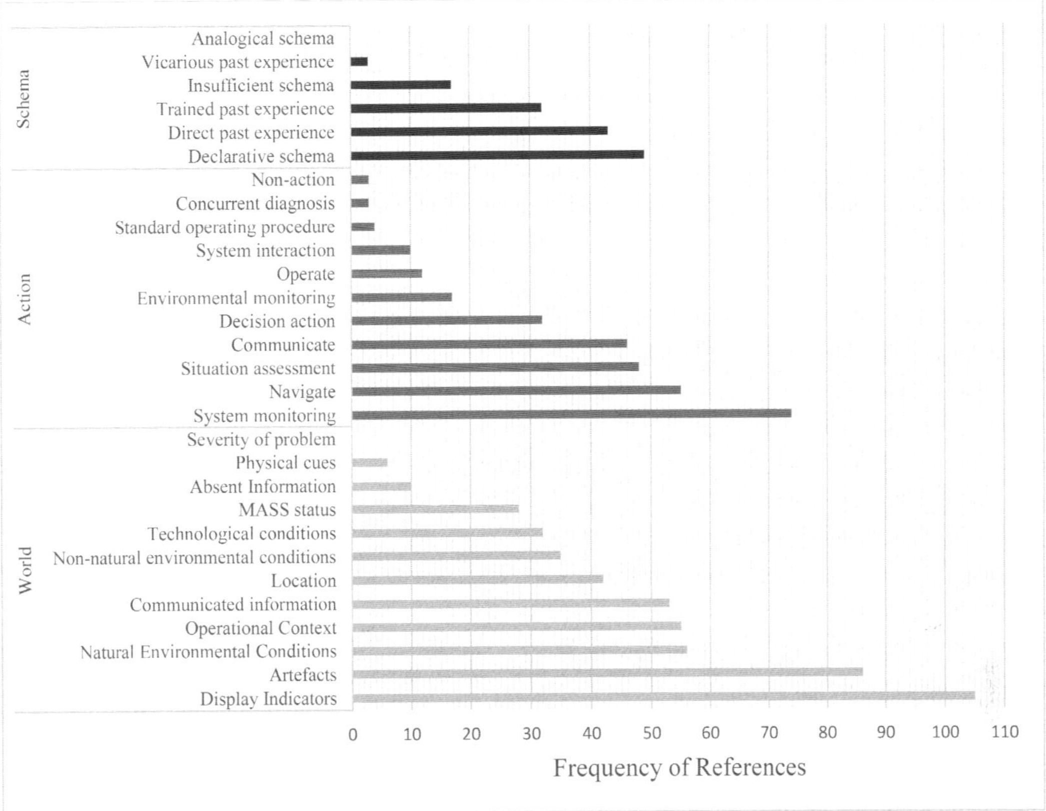

Figure 2: Frequency of reference for each SAW code in the interview transcripts.

3.2 Design Principles

Design principles were then generated from the SAW sub-types and associated transcript data. Next, the design principles were grouped based on the operator task they supported and displayed on the PCM framework to show the considerations for the different sub-types of the PCM to support decision-making. Four key operator tasks were identified in the design principles by iteratively reviewing the transcript data from the SAW sub-types. Firstly, aiding the operator in developing expectations of the automated systems' behaviours (Figure 3) and then supporting the operator in supervising the automated system, particularly around the navigation of the MASS (Figure 4). The other tasks were aiding the operator's ability to operate MASS uncrewed and remotely from a RCC (Figure 5).

3.2 (a) Developing expectations of the automation

For operators to be able to work within a human-machine team they require a good understanding of what they can expect from the automated systems, to be able to anticipate its behaviour. The design principles for supporting the development of appropriate expectations of the automation are shown in Figure 3, showing the considerations to support schema development, the types of world information needed and the actions that will need to be supported.

For the world information to understand the automated system operators will need clear indications of key system information such as the automation mode, the health status and system parameters (e.g., speed and heading), to be able to determine whether the information meets their expectations. If key information such as the mode of the automated system is not clear and/or operators do not receive sufficient feedback from the system it could cause mode confusion, where the operator thinks the automation is in an alternative mode and especially in cases where there are multiple levels of automation for the system [29, 30]. Therefore, system transparency will need to be considered, so operators can understand what the system is doing to be able to determine whether it matches their expectations of the automation [31].

To support the schema component of the PCM, it will be necessary to ensure that operators have time and experience with the system to develop an appropriate schema of operations, to aid their ability to interpret the information from the automated system and understand what actions to take. It has been shown that having prior experience with an automated system can influence acceptance of a system including systems using higher levels of automation [32].

For the operator's actions, the handover of control between the operator and the automation will be a key part of how effectively the human-machine team can operate. Training operators in how the automation works and what they should expect from the automation will play an important factor in maintaining the safety of the overall system. Similarly, it has been highlighted that there is a need to consider the training of MASS operators for difficult high-stress situations such as unanticipated handovers from the system and, that current training and guidance do not include the role of human users and interventions when working with automated systems [33]. Training on the automated system's limitations and strengths will allow operators to develop appropriate mental models of the system and calibrate their trust [33, 34]. However, a greater emphasis on the automation's strengths can lead to automation complacency and overconfidence in the system [35]. Demonstration-based training can be used to give operators a more detailed knowledge of the automated system's decision-making and event-based training in a simulator would allow operators to gain experience interacting with the system [36].

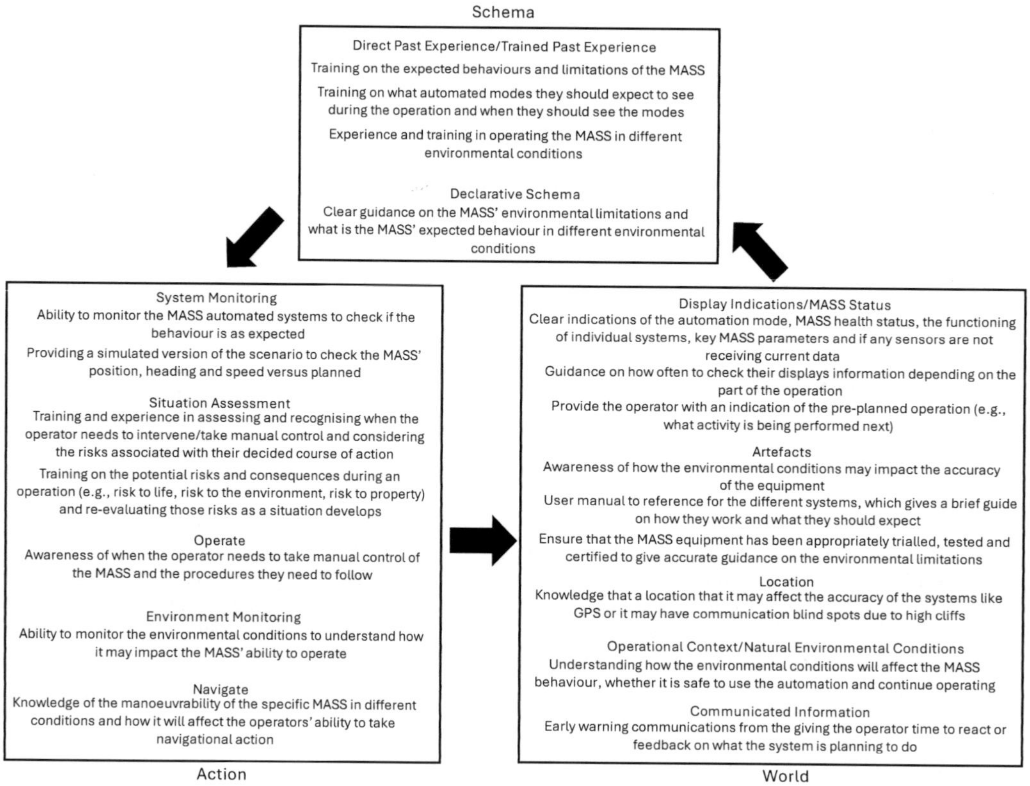

Figure 3: Design principles to aid operators in developing appropriate expectations of the automated systems.

3.2 (b) Supporting the operator's supervisory role

A MASS operator's role will be predominantly a supervisory role, where the operator will be monitoring the system, to provide a backup to the automated system and intervene as needed. The design principles and considerations for supporting the supervisory role are shown in Figure 4. It will be necessary to consider how the operators' tasks will be designed to support them in being able to monitor the systems for long periods, cognitive underload has already been seen in the operation of conventional vessels [37]. The introduction of higher levels of automation may make this effect greater for MASS operators and it could reduce an operator's takeover performance, their perception and situational awareness [33, 38]. Cognitive underload can reduce performance when the workload transitions from a low level to a high level when the operator needs to intervene [39, 40]. In automated driving, it has been suggested that giving the operator secondary tasks could reduce this effect by increasing their cognitive load and keeping them better engaged [38, 41, 42]. Adaptive automation could be used to overcome cognitive underload, where the system can give manual control to the operator or a lower level of control to reengage them in the task or systems could be designed to adapt to the operator's state a give the operator tasks to manage [43, 44].

Increasing the system transparency has been suggested for MASS to help the operator monitor the automated system, to allow the operator to collect information on the system and evaluate its functioning, and supervise it taking collision and grounding avoidance actions [9, 45]. Using animations or a trial function has been suggested to increase transparency in automated collision avoidance manoeuvres, to show the predicted outcome of the manoeuvre and to increase understanding of what the MASS is planning to do [46]. The identification of the target ships, the COLREG rule being followed and a reason for why the system had decided on the manoeuvre (e.g., type of ship, aspect) could also increase an operator's understanding [46]. Increasing the transparency of a collision and grounding avoidance system has been shown to improve situational awareness, although no effect between transparency and cognitive workload was found [13].

Setting navigation routes and updating system configurations for a navigation route has been identified as a risky task in MASS operations due to their complexity and the risks of collisions with other vessels [47]. Therefore MASS operators will need advanced navigational training [47], which has also been found in the schema design principles shown in Figure 4. It has been shown that having onboard experience in navigating a vessel can improve a novice operator's navigational ability versus an operator with no onboard experience [48]. The findings have also highlighted that operators will need specific experience with MASS, particularly in situations where conventional vessels react unexpectedly to them as they are not familiar with MASS leading to potentially dangerous and confusing situations.

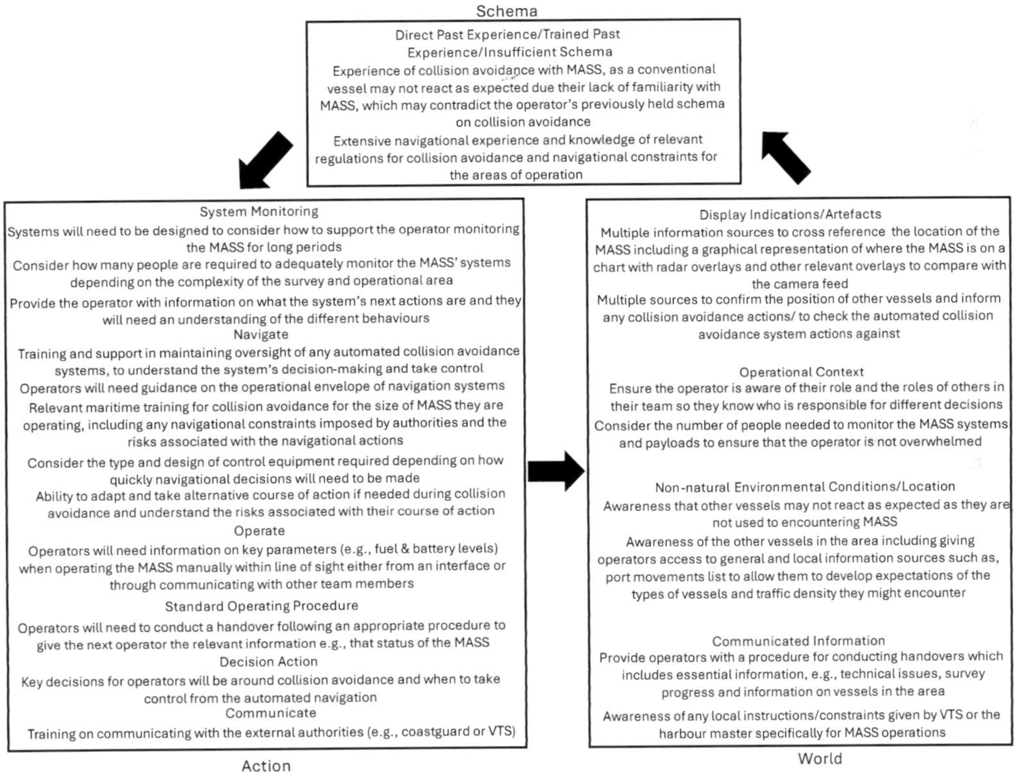

Figure 4: Design principles to support operators in monitoring the automated systems, particularly the navigational systems.

3.2 (c) Supporting uncrewed operations

Figure 5 shows the design principles that were generated from the SAW subtypes for operations from an RCC, including the additional world information that will be needed and supporting operators in performing actions remotely. When operating MASS uncrewed one challenge is the lack of physical cues to interpret when making decisions. Kristoffersen [31] recommended engaging with users that have experience from conventional vessels to ensure that the information they are provided with is sufficient to replace their sensory information and that they still have high levels of situation awareness. It will be important to consider how this additional information is given to the operator as current user interfaces predominantly rely on visual information, which could cognitively overload the operator [31]. 3D and virtual reality human-machine interfaces could be used to help overcome the issues associated with remote operations and the design principle for display indications in providing information to the operator to replace their physical cues, although there are various human factors considerations which need to be investigated further [49, 50].

The usability of the interface and the types of controls selected for the RCC will be an important factor, especially if the MASS is being operated on a host ship. The usability of the operator's interface is a key consideration as it can affect the operator's situational awareness and workload [51]. Conventional navigational instruments were given low usability scores when they were used in an RCC rather than on a conventional bridge, suggesting that interfaces will need to be redesigned for RCCs to reflect the differences in how the system is being operated and support the operators' ability to gather information [2, 52].

Operators will need experience in using alternative sources of information from the technological systems to maintain their situational awareness and be able to make effective decisions. Due to this difference in the type of information they will use, operators will need training to practice how they will gather the information and to develop new schema [31]. Developing new schema for remote operations may be difficult for them due to operators' heavily learnt schema from their mariner training, and experience in how they would normally verify information from technological systems with physical cues before making a decision. A study has investigated operators using the same information/instruments to navigate when operating remotely but found it was more challenging for the operators, as verification was more difficult and they had to use additional tools (e.g., camera feed and radar) to compensate [2].

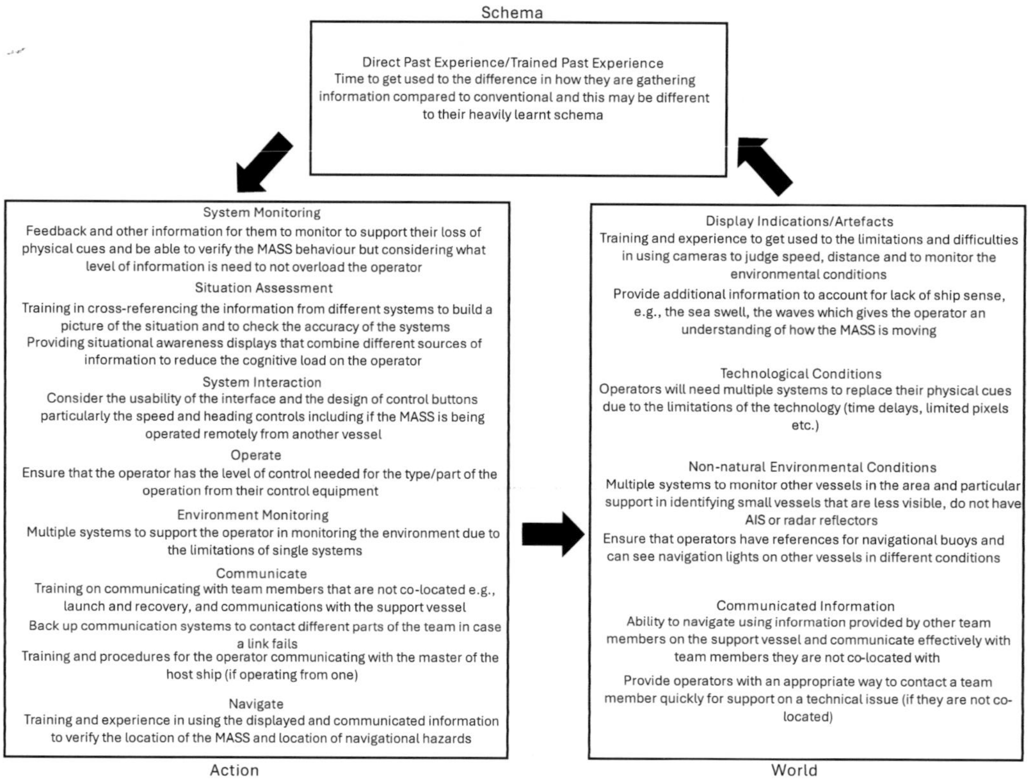

Figure 5: Design principles for supporting operators in the operator of uncrewed MASS from RCCs.

4. DISCUSSION

SWARM interviews with MASS operators and SMEs have been used to develop design principles to consider for optimising operators' decision-making. The design principles support the three components of the PCM framework, schema, action and world, across four key tasks in MASS operation. The key tasks identified were evaluating the automated system's behaviour against their expectation; supervising the automated systems, and operating remotely without physical cues.

For developing appropriate automation expectations, the design principles were around the world information, display indications, artefacts and the environmental conditions that the operator can utilise to evaluate whether the information matches their expectations, or whether they need to take action and intervene. Therefore, it will be necessary to consider the system's transparency to aid understanding and reduce the potential for mode confusion [29-31, 58]. Additionally, it will be important to consider the operator's training requirements to support the development of appropriate schema when interacting with the automation [33, 34]. For optimising the operator in supervising the automated systems and especially the navigation the design principles considered factors such as cognitive underload [38-40] and the system's transparency as operators will need relevant information to understand and evaluate the system's reasoning [9, 45]. For operating without physical cues design principles were suggested including the need for multiple systems to provide further information to aid operators in maintaining their situational awareness [2, 31] and to consider the usability of the system on how that information additional might be displayed [2, 52].

The findings further demonstrate the SWARM can be used to generate user-centred design recommendations and the method applies to other safety-critical domains such as MASS in addition to aviation, only minor adjustments were made to the SAW codes for analysis. Other applications of SWARM have investigated aeronautical decision-making, including design recommendations for an engine monitoring tool for a flight deck for bird strikes, a system for engine oil leaks and the operation of multiple uncrewed aerial vehicles [23-25]. These applications of SWARM were also used to identify key operator tasks and generate recommendations to support them [23-25]. It has been highlighted that engaging with users will be important in the design and development of future MASS systems to ensure operators have the necessary information and systems are designed to optimise their decision-making [31, 59].

Due to the infancy of MASS, a small sample size was used and some of the participants did not have operational experience, although they had extensive knowledge of how MASS are and will be operated in the future. Future work could investigate the decision-making of a larger sample of operators to explore more challenges that have been experienced in MASS operations including any future critical incidents to understand why they occurred and how they were recovered. A limitation of the SWARM interview technique is the challenge in eliciting schema data, which could explain the lower number of schema references as mental models are often used unconsciously, so participants may not be able to articulate schema data as easily as both world and action information [60, 61]. Further work could investigate operator's decision-making in real-time by using verbal protocol analysis with operators who are operating in a real-world or simulated scenario, as there are also limitations in using recall methods. However, the decision-making design principles and incident models developed here have identified initial areas that need to be considered during the design of MASS systems and it will be necessary to understand how these principles can be applied through further research.

5. CONCLUSIONS

Design principles have been developed to support decision-making in the operation of MASS from RCCs by using the SWARM interview technique. The PCM framework has been used to explore MASS operators' decision-making to investigate the world information used by operators, how their mental models affect decision-making and their key decision actions. The results showed that key challenges will be operators having appropriate expectations of the automation, supporting operators in monitoring the highly automated systems effectively and replacing their lack of physical senses. Further research should explore how these design principles could be applied in the design of future MASS systems.

6. ACKNOWLEDGEMENTS

This work was conducted as part of an Industrial Cooperative Awards in Science & Technology in collaboration with Thales. Any views expressed are those of the authors and do not necessarily represent those of the funding body.

7. REFERENCES

1. International Maritime Organisation. Available from: https://www.imo.org/en, 2023

2. MAN, Y., et al., Human factor issues during remote ship monitoring tasks: An ecological lesson for system design in a distributed context. *International Journal of Industrial Ergonomics*, 2018. 68: p. 231-244.

3. HAN, C., et al., Future Directions for Human-Centered Transparent Systems for Engine Room Monitoring in Shore Control Centers. *Journal of Marine Science and Engineering*, 2022. 10(1): p. 22.

4. STATECZNY, A. and P. BURDZIAKOWSKI, Universal autonomous control and management system for multipurpose unmanned surface vessel. *Polish Maritime Research*, 2019. 26(1): p. 30-39.

5. WAZAZ, H., et al. Use of a low-cost unmanned surface vessel for bathymetricy surveys and pocket beach sediment dynamics monitoring. *in IEEE International Workshop on Metrology for the Sea Learning to Measure Sea Health Parameters (MetroSea)*. 2022. Milazzo, Italy.

6. LI, J., et al., A survey of maritime unmanned search system: Theory, applications and future directions. *Ocean Engineering*, 2023. 285: p. 115359.

7. CASALINO, G., TURETTA, A., and SIMETTI, E.,. A three-layered architecture for real time path planning and obstacle avoidance for surveillance USVs operating in harbour fields. in OCEANS 2009-EUROPE. 2009.

8. MADSEN, A. and T.-e. KIM, A state-of-the-art review of AI decision transparency for autonomous shipping. 2024. 8.

9. RAMOS, M., Utne, I., and Mosleh A., Collision avoidance on maritime autonomous surface ships: Operators' tasks and human failure events. *Safety Science*, 2019. 116: p. 33-44.

10. PORATHE, T., PRISON, J., and MAN, Y., Situation awareness in remote control centres for unmanned ships. 2014.

11. ENDSLEY, M. and JONES, W., Situation Awareness Information Dominance & Information Warfare. 1997: p. 94.

12. CHAUVIN, C., CLOSTERMANN, J.-P., and HOC, J.-M., Situation Awareness and the Decision-Making Process in a Dynamic Situation: Avoiding Collisions at Sea. *Journal of Cognitive Engineering and Decision Making*, 2008. 2: p. 1-23.

13. VAN DE MERWE, K., et al., The Influence of Agent Transparency and Complexity on Situation Awareness, Mental Workload, and Task Performance. *Journal of Cognitive Engineering and Decision Making*, 2024. 18(2): p. 156-184.

14. WARM, J.S., PARASURAMAN, R., and MATTHEWS, G., Vigilance Requires Hard Mental Work and Is Stressful. *Human Factors*, 2008. 50(3): p. 433-441.

15. ENDSLEY, M. and KIRIS, E., The Out-of-the-Loop Performance Problem and Level of Control in Automation. *Human Factors: The Journal of the Human Factors and Ergonomics Society*, 1995. 37.

16. PARASURAMAN, R. and RILEY, V., Humans and Automation: Use, Misuse, Disuse, Abuse. *Human Factors*, 1997. 39(2): p. 230-253.

17. RØDSETH, Ø.J., LIEN WENNERSBERG, L, A., and NORDAHL, H., Towards approval of autonomous ship systems by their operational envelope. *Journal of Marine Science and Technology*, 2022. 27(1): p. 67-76.

18. MACKINNON, S.N., et al. Command and control of unmanned vessels: Keeping shore-based operators in-the-loop. in *18th International Conference on Ships and Shipping Research*, NAV 2015. 2015.

19. VEITCH, E., et al., Collaborative Work with Highly Automated Marine Navigation Systems. *Computer Supported Cooperative Work: CSCW: An International Journal,* 2024. 33(1): p. 7-38.

20. RØDSETH, Ø.J., et al. Communication architecture for an unmanned merchant ship. in 2013 *MTS/IEEE OCEANS* - Bergen. 2013.

21. ORASANU, J. and CONNOLLY, T., The reinvention of decision making, in *Decision making in action: Models and methods*. 1993, Ablex Publishing: Westport, CT, US. p. 3-20.

22. PLANT, K L,. and STANTON, N A., The development of the Schema World Action Research Method (SWARM) for the elicitation of perceptual cycle data. *Theoretical Issues in Ergonomics Science*, 2016. 17(4): p. 376-401.

23. BANKS, V.A., et al., Using the Perceptual Cycle Model and Schema World Action Research Method to generate design requirements for new avionic systems. *Human Factors and Ergonomics in Manufacturing & Service Industries*, 2021. 31(1): p. 66-75.

24. PARNELL, K.J., et al., Generating Design Requirements for Flight Deck Applications: Applying the Perceptual Cycle Model to Engine Failures on Take-off. *International Journal of Human–Computer Interaction*, 2021. 37(7): p. 611-629.

25. PARNELL, K.J., et al., Trustworthy UAV Relationships: Applying the Schema Action World Taxonomy to UAVs and UAV Swarm Operations. *International Journal of Human–Computer Interaction*, 2022: p. 1-17.

26. LYNCH, K.M., et al., Maritime autonomous surface ships: can we learn from unmanned aerial vehicle incidents using the perceptual cycle model? *Ergonomics*, 2023. 66(6): p. 772-790.

27. BANKS, V.A., PLANT, K L., and Stanton, N A., Driver error or designer error: Using the Perceptual Cycle Model to explore the circumstances surrounding the fatal Tesla crash on 7th May 2016. *Safety Science*, 2018. 108: p. 278-285.

28. PLANT, K L. and STANTON, N A., The development of the Schema-Action-World (SAW) taxonomy for understanding decision making in aeronautical critical incidents. *Safety Science*, 2016. 99: p. 23-35.

29. ENDSLEY, M.R., Autonomous Driving Systems: A Preliminary Naturalistic Study of the Tesla Model S. *Journal of Cognitive Engineering and Decision Making*, 2017. 11(3): p. 225-238.

30. SILVA, S.S. and HANSMAN, R J., Divergence Between Flight Crew Mental Model and Aircraft System State in Auto-Throttle Mode Confusion Accident and Incident Cases. *Journal of Cognitive Engineering and Decision Making*, 2015. 9(4): p. 312-328.

31. KRISTOFFERSEN, C., Unmanned autonomous vessels and the necessity of human-centred design. *DS 101: Proceedings of NordDesign* 2020, Lyngby, Denmark, 12th-14th August 2020, 2020: p. 1-12.

32. BEKIER, M. and MOLESWORTH, B R C., Altering user' acceptance of automation through prior automation exposure. *Ergonomics*, 2017. 60(6): p. 745-753.

33. TAM, K., et al., The potential mental health effects of remote control in an autonomous maritime world. *Journal of International Maritime Safety, Environmental Affairs, and Shipping*, 2021. 5(2): p. 40-55.

34. DREXLER, D.A., et al. Handover Process Models of Autonomous Cars Up to Level 3 Autonomy. in *2018 IEEE 18th International Symposium on Computational Intelligence and Informatics* (CINTI). 2018.

35. SINGER, J., et al., Driver Expectations of a Partial Driving Automation System in Relation to Branding and Training. *Human Factors*, 2022. 66(5): p. 1531-1544.

36. FLEMING, E. and PRITCHETT, A., SRK as a framework for the development of training for effective interaction with multi-level automation. *Cognition, Technology & Work*, 2016. 18(3): p. 511-528.

37. CHAM, B.S., et al., Investigating the joint effects of overload and underload on chronic fatigue and wellbeing. *Work & Stress*, 2021. 35(4): p. 344-357.

38. CHENG, T., et al., Analysis of human errors in human-autonomy collaboration in autonomous ships operations through shore control experimental data. *Reliability Engineering & System Safety*, 2024. 246: p. 110080.

39. YOUNG, M.S., et al., State of science: mental workload in ergonomics. *Ergonomics*, 2015. 58(1): p. 1-17.

40. YOUNG, M.S. In Search of the Redline: Perspectives on Mental Workload and the 'Underload Problem'. 2021. *Cham: Springer International Publishing.*

41. GOLD, C., HAPPEE, R., and BENGLER, K., Modeling take-over performance in level 3 conditionally automated vehicles. *Accident Analysis & Prevention*, 2018. 116: p. 3-13.

42. MISHLER, S. and CHEN, J., Boring But Demanding: Using Secondary Tasks to Counter the Driver Vigilance Decrement for Partially Automated Driving. *Human Factors*, 2024. 66(6): p. 1798-1811.

43. BYRNE, E.A. and PARASURAMAN, R., Psychophysiology and adaptive automation. *Biological Psychology*, 1996. 42(3): p. 249-268.

44. DEGUZMAN, C.A., KANAAN, D., and DONMEZ, B., Attentive User Interfaces: Adaptive Interfaces that Monitor and Manage Driver Attention, in User Experience Design in the Era of Automated Driving, A. Riener, M. Jeon, and I. Alvarez, Editors. 2022, *Springer International Publishing: Cham*. p. 305-334.

45. VAN DE MERWE, K., MALLAM, S., and NAZIR, S., Agent Transparency, Situation Awareness, Mental Workload, and Operator Performance: A Systematic Literature Review. *Human Factors*, 2022. 66(1): p. 180-208.

46. MADSEN, A.N., BRANDSÆTER, A., and AARSET, M V., Decision Transparency for enhanced human-machine collaboration for autonomous ships. *Human Factors in Robots, Drones and Unmanned Systems*, 2023.

47. LIU, J., et al., Prediction of human–machine interface (HMI) operational errors for maritime autonomous surface ships (MASS). *Journal of Marine Science and Technology*, 2022. 27(1): p. 293-306.

48. HWANG, T. and YOUN, I -H., Navigation Scenario Permutation Model for Training of Maritime Autonomous Surface Ship Remote Operators. *Applied Sciences*, 2022. 12(3): p. 1651.

49. LAGER, M., TOPP, E A., and MALEC J., Remote Supervision of an Unmanned Surface Vessel - A Comparison of Interfaces. in *2019 14th ACM/IEEE International Conference on Human-Robot Interaction* (HRI). 2019.

50. KAZEMI, R. and LEE, S C., Human Factors/Ergonomics (HFE) Evaluation in the Virtual Reality Environment: A Systematic Review. *International Journal of Human–Computer Interaction*, 2023: p. 1-17.

51. NIELSEN, J., Usability engineering. 1994: Morgan Kaufmann.

52. KARVONEN, H. and MARTIO J., Human Factors Issues in Maritime Autonomous Surface Ship Systems Development. in *Proceedings of the 1st International Conference on Maritime Autonomous Surface Ships*. 2019. SINTEF Academic Press.

53. MEHROTRA, S., et al., Human-Machine Interfaces and Vehicle Automation: A Review of the Literature. Accident Analysis & Prevention, 2022. 109: p. 18-28.

54. WICKENS, C.D., et al., Complacency and Automation Bias in the Use of Imperfect Automation. *Human Factors*, 2015. 57(5): p. 728-739.

55. BAHNER, J E., HÜPER, A -D., and MANZEY, D., Misuse of automated decision aids: Complacency, automation bias and the impact of training experience. *International Journal of Human-Computer Studies*, 2008. 66(9): p. 688-699.

56. HAN, C., et al., Situation awareness for engine room monitoring at future shore control centre. *Australian Journal of Maritime and Ocean Affairs*, 2024.

57. STRAUCH, B., The Automation-by-Expertise-by-Training Interaction: Why Automation-Related Accidents Continue to Occur in Sociotechnical Systems. *Human Factors*, 2017. 59(2): p. 204-228.

58. VEITCH, E. and ALSOS, O A., Human-Centered Explainable Artificial Intelligence for Marine Autonomous Surface Vehicles. *Journal of Marine Science and Engineering*, 2021. 9(11): p. 1227.

59. VU, V D. and LÜTZHÖFT, M., Improving human-centred design application in the maritime industry – challenges and opportunities. *International Conference on Human Factors* 2020, 2020.

60. PLANT, K L. and STANTON, N A., The process of processing: exploring the validity of Neisser's perceptual cycle model with accounts from critical decision-making in the cockpit. *Ergonomics*, 2015. 58(6): p. 909-923.

61. PLANT, K L. and STANTON, N A., The explanatory power of Schema Theory: theoretical foundations and future applications in *Ergonomics. Ergonomics*, 2013. 56(1): p. 1-15.

8. AUTHORS' BIOGRAPHY

Kirsty Lynch is a PhD Student in Human Factors Engineering within the Transportation Research Group of the Faculty of Engineering and Physical Sciences, University of Southampton and is a Human Factors Engineer at Thales UK. Kirsty graduated with a MEng (Hons) in Mechanical Engineering with Automotive from the University of Southampton in 2019. Kirsty's PhD is investigating decision-making in human-machine teams, to explore how systems can be designed to support accountable decision-making at remote control centres. The human-machine team that is the focus of the PhD is human operators and maritime autonomous surface ships. Her main research interest centres around investigating how human-machine interfaces can be designed to best support operators' decision-making when working with unmanned systems.

Professor Mark Young is Professor of Human Factors in Transport within the Transportation Research Group at the University of Southampton and current President (2024-25) of the Chartered Institute of Ergonomics and Human Factors (CIEHF). Mark has nearly 30 years' experience working in human factors across transport modes in both academia and industry, and has written over 70 peer-reviewed journal papers and five books. He is a Chartered Ergonomist and a Fellow of the CIEHF.

Dr Dominic Taunton is an Associate Professor in Ship Science and Maritime Engineering within Engineering and Physical Sciences at the University of Southampton. He is the Director of Programmes for Maritime Engineering. His research interests lie in improving the design and operation of ships and other maritime artefacts, in particular high speed craft such as semi-displacement catamarans and hard chine planning craft, through the development of accurate but efficient tools for predicting and evaluating their performance. His other interests have explored the Human Element, this research developed from a project to integrate human performance into the design of high speed planning craft. This integration of the human element into the design process has resulted in collaborations with human element experts in order to improve ship design and operation. This approach has led to the development of human like autonomy behaviours to improve the seakeeping and fuel consumption of autonomous surface vehicles.

Dr Aaron Roberts received the B.Sc. degree in Psychology from the University of Liverpool, Liverpool, UK, in 2007 and a PhD in Applied Cognitive Psychology in 2012 from the same institution. He is also a Chartered Psychologist and a Chartered Human Factors practitioner. Aaron was previously a Senior Research Fellow in Human Factors at the University of Southampton, Southampton, UK. During this period Aaron led a number of research projects focused on evaluating defence mission systems from a sociotechnical systems perspective. Aaron has remained a visiting Fellow at the University of Southampton but is currently a Senior Operability Expert within Thales UK. Aaron's research interest's centre around the cognitive capacities and physiology of humans, in particular what changes occur during situations of increased demand. Aaron has evaluated the cognitive function, teamwork processes and sociotechnical system performance of paramedics, pilots, submariners, fire service personnel, police officers and nurses in simulated operational environments.

Dr Victoria Banks, also known as Victoria Steane, is a Human Factors practitioner with over 10 years' experience working in multiple transport domains (automotive, aviation, maritime). She holds a BSc in Psychology and EngD from the University of Southampton. She currently works within Thales UK as a Human Factors Researcher but is also a Visiting Research Fellow at the University of Southampton and a Research Associate at the University of Bristol. Her main research interests centre on agent-based modelling of complex systems operating at varying levels of autonomy and the design and development of decision support aids and intervention strategies to optimise human performance.

Dr Katherine Plant is an Associate Professor in Human Factors Engineering in the Transportation Research Group in the Faculty of Engineering and Physical Sciences at the University of Southampton, UK. She holds a BSc (Hons) degree in Psychology from the University of Bath (2008) and a PhD in Human Factors Engineering from the University of Southampton (2015). She is also a Chartered member of the Chartered Institute for Ergonomics and Human Factors. Katherine's main research interests centre around naturalistic decision making in complex sociotechnical systems, particularly the influence that the interaction of the environment and past experiences have on these decisions. Katherine manages a research portfolio across a range of transportation domains including aviation, road safety, autonomous systems and maritime, which includes the supervision of nine PhD students. In 2016 Katherine authored the book 'Distributed Cognition and Reality: How pilots and crews make decisions' (CRC Press) and this body of research was awarded the Air Pilots Trust award for Aviation Safety Research (2014).

Human Factors 2024
8th – 9th October 2024
Wageningen, Netherlands

REALIZING LARGER CLICK SURFACES FOR MARITIME USER INTERFACES USING FUNCTION GROUPING

K Nordby and **J E Fauske,** Oslo School of Architecture and Design, Norway

SUMMARY

Maritime regulation establishes a minimum button size of 15 mm for most maritime digital interfaces with touch screens. However, there's a contention that this minimum size might be too small for efficient operation in challenging maritime environments. Notwithstanding these concerns, many interfaces continue to adopt the 15mm size.

In this study, we introduce a novel design concept that notably amplifies click surface areas. This enhancement is achieved by clustering multiple functions within more extensive zones and by refining sub-menus to improve readability, thereby offering expansive click surfaces for all integrated functions within a singular view. This reimagined structure diverges from conventional maritime interface designs, optimizing user accuracy in targeting specific surface areas. Additionally, our method streamlines keystroke interactions, enabling users to traverse the interface with fewer clicks and presenting more generous click surfaces suitable for mouse or trackball use.

We provide a comprehensive description of our proposed concept and evaluate it using Fitts' law. Furthermore, we critically examine the potential shortcomings and advantages of our design proposal. In summary, we argue that our design concept likely paves the way for reduced errors and streamlined interactions, particularly beneficial for users experiencing situational impairments at sea.

1. INTRODUCTION

Touch screens are increasingly being installed in maritime workplaces, offering numerous benefits over physical buttons, such as lower cost, flexibility, space efficiency, and ease of development. Despite these advantages, their prevalence in maritime operational environments, such as ship bridges, can be questioned. Maritime workplaces are harsh environments where users operate equipment under constantly changing conditions. Factors like fatigue, heave motion due to waves, and varying light conditions can temporarily impair users [1], making touch screens exceedingly difficult to operate. Furthermore, since the primary activity is maintaining a view of the outside [2,3], using devices that require direct coordination between eyesight and interaction can be problematic in high-demand situations.

Since touch screens are already common in maritime workplaces, this work turns to ongoing efforts to improve touch screen-related practices. Specifically, we report on efforts to realize large click surfaces in maritime equipment. Maritime equipment is often information-dense, and regulations often mandate information that needs to always be on-screen. Because of this, equipment producers often find that they need to reduce the size of screen content and stack it densely to meet functional requirements. Currently, the minimum requirement for touch areas for maritime interfaces is 15x15 mm, regardless of the type of system or typical use context [4]. This is larger than the minimum touch area often used for public digital services, where WCAG 2.1 recommends a 12 mm minimum touch surface. WCAG is a commonly used framework supporting accessibility in digital interfaces [5].

While the maritime standard is considerably larger than what is mandated in the WCAG guideline, we believe that 15 mm might still be too small in many maritime use cases, given the environmental challenges on a ship, motion of the environment, and the need to maintain gaze away from the screen to ensure situational awareness of the outside view.

Our study addresses the challenge of limited touch areas on maritime touch screens by focusing on the interaction design of touch interfaces. Specifically, we demonstrate an alternative design strategy for touch interaction in a specific design case. We argue that we can leverage the dynamic possibilities of digital user interfaces to improve click surfaces without affecting information density on a screen. We demonstrate this by showing a design concept that groups similar functions into larger click areas. We address this as function clustering and show how the concept can significantly increase the possible touch area for an application without reducing the content that can be shown on screen. We compare the design concept with a traditional concept using 15 mm minimum buttons.

The design case presented here has been developed in the OpenAR research project funded by the Norwegian Research Council [6]. The case builds on and informs the development of the OpenBridge design system, an open-source design guideline for maritime user interfaces [7]. This is a preliminary study aimed at building an initial assessment of the new interaction concept prior to formal usability studies with users.

2. THE DESIGN CASES

Our design case is based on two instances of a Conning user interface for ships bridges. Conning is a typical maritime interface that gathers a range of data critical for operating a ship. It typically contains central information about navigation, manoeuvring, propulsion systems, and other relevant data. We present two slightly different design examples (Figure 1) focusing on the list in the top left corner in both interfaces that shows various ship data. The list connects the data (such as heading, course over ground, and rate of turn) to the sensors delivering it, allowing users to see sensor status and change the sensors used for each data item.

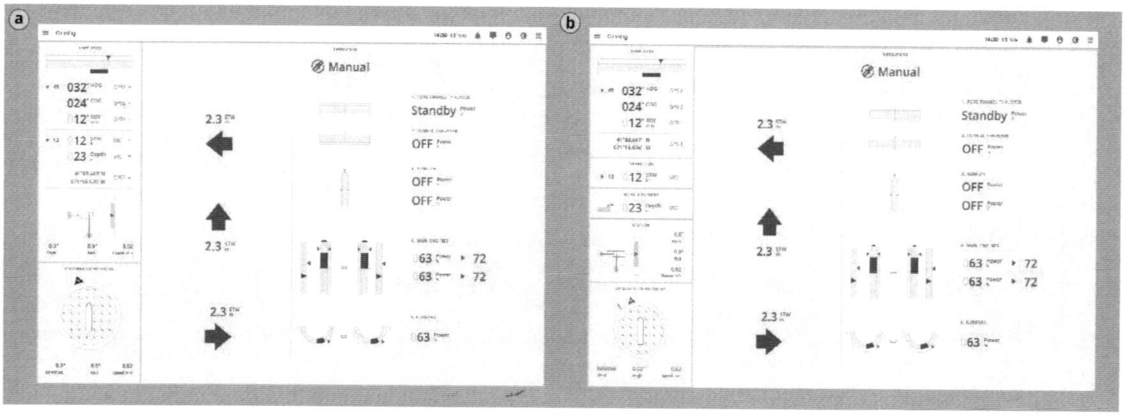

Figure 1. Conning design cases. UI 1 (a) with small buttons and UI 2 (b) with large buttons

2.1 TRADITIONAL INTERFACE DESIGN (UI 1)

The first interface (UI 1), seen in Figure 2, includes a separate sensor button in each data line. The buttons adhere to the 15 mm minimum size requirement. This interface uses a standard user interface data readout component from the OpenBridge design system library. When the sensor button is pressed, it opens a secondary dropdown menu near the button. An arrow on the right side of the button signifies the dropdown functionality, following common user interface conventions.

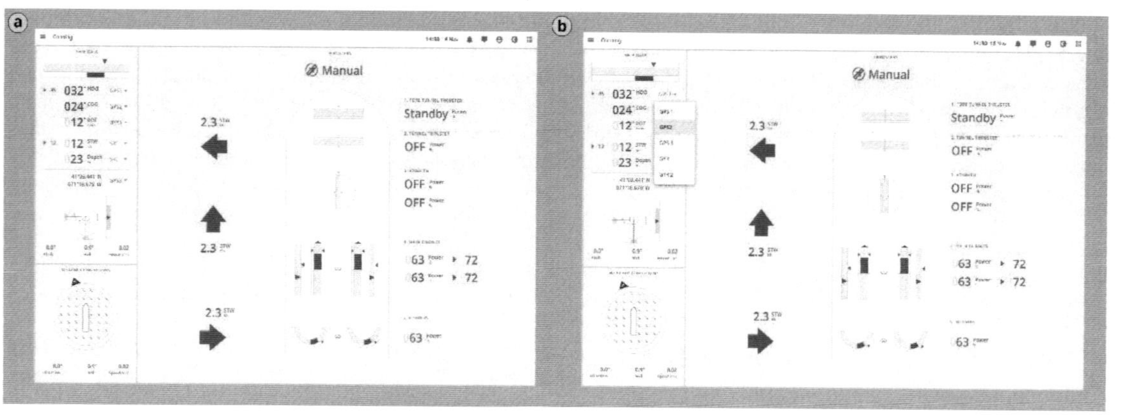

Figure 2. UI1 main page (a), UI1 with dropdown menu open (b).

2.2 NEW INTERFACE DESIGN (UI 2)

The second interface (UI 2), seen in Figure 3, is almost identical to the first interface at the base level. However, there are some slight changes. In this example, the individual dropdown menus are removed, and the entire sensor group is integrated into a single, very large button. This function clustering allows us to significantly increase the button size from the 15 mm minimum to over 100 mm. In our example, we have organized the sensors into three groups, with speed through water and depth having individual large buttons as well.

The new design reduces the need for graphics signifying interaction for separate buttons. This frees up horizontal space and reduces some clutter. However, clustering the buttons on the main page results in a need to display much more options in the popup menu. We address this challenge by increasing the size of the secondary menu. As a temporary interface, we are not required to keep this small. Thus, we can use larger buttons here as well and display all sensor options on a single page. By making the menu very large, we can show relevant data in the same view as the buttons. Additionally, since the buttons are related, it is possible to see the selected state and available options of all related equipment simultaneously. Importantly, all the functions are still available as *simple operator action* using two clicks [8] even if the content has been reorganized in UI 2.

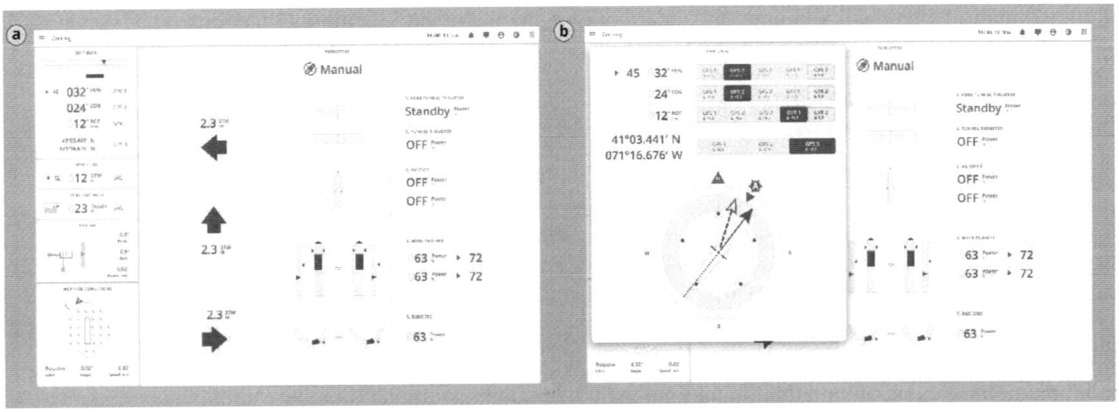

Figure 3. UI2 main page (a), UI2 with large menu open (b).

3. TRAVEL DISTANCE AND BUTTON SIZE

To compare the design proposals, we performed a simple analysis of the interaction involved by evaluating the travel distance versus the button size, using Fitts' Law as a basis. Fitts' Law is a common framework used in human-computer interaction to predict the time it takes for a pointer, such as a mouse cursor or a finger, to move to a target to interact with it [9]. Fitts' Law presents an equation that establishes the relationship between travel distance and the size of interaction targets. In short, the model framework shows that the longer the distance a pointer moves towards a target, the longer it takes. However, it also shows that the larger the target size, the shorter the time it will take to move to the target. In this article we will just use the relation between travel distance and button size as a general guide to make an initial analysis.

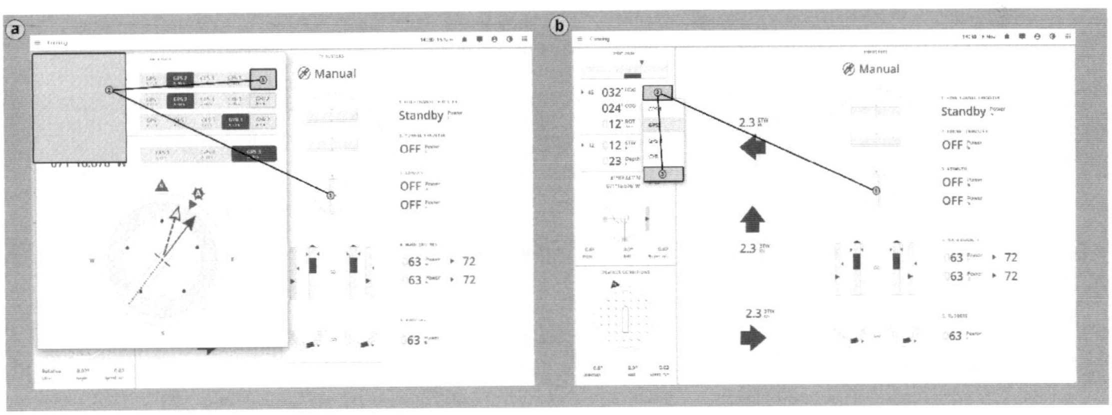

Figure 4. UI 2 pointer travel (a). UI 1 pointer travel (b).

In Figure 4, we visualized a finger starting from the same point (1), moving to the button area and pressing it in the base layout (2), and then moving to the sensor setting (3). The grey areas indicate the button size, and the lines represent the approximate travel distance.

In Figure 4, we can see that the travel distance for the first step (1-2) is similar between the two interfaces. However, in the next step (2-3), we see that the travel distance between the buttons farthest apart in UI 1 (11 cm) is shorter than in UI 2 (15). This travel distance is countered by the button sizes. In the first step, there is a significant difference in button size. According to Kenzie and Buxton's suggestion to use the smallest distance in a rectangular button in the model [10], UI 1 has 15 mm buttons while UI 2 has 100 mm buttons. This suggests that UI 2 is much more efficient in the first step. In the second step, the differences are smaller, with UI 1 having 15 mm buttons and UI 2 having 20 mm buttons. Given this difference, Fitts law predicts that we can assume that the second step is likely to be the same speed for both interfaces since the button size compensate for the added travel distance.

This is a simplified assessment, and several other factors affect the interaction. For instance, the first large button in UI 2 is placed along a screen edge, effectively making the target surface much larger since it reduces the chance of missing the target. Furthermore, our assumption uses the real click area, including clickable padding, rather than the visual click area. This is slightly inaccurate since the appearance of smaller buttons might reduce click speed, even if the error rate is reduced. Further, we do not assess the potential error rate in this assessment.

Other effects

In addition to having larger button areas, the new interface changes the structure of interaction in the interface. The number of clicks to access a single function is identical to previous systems. However, if users are adjusting multiple settings at once, the number of clicks is significantly reduced by removing the need to open a new dropdown for each instrument (Figure 5,1). As such, it has the potential to increase the efficiency of the interface for some tasks.

Further, the new interface replaces traditional dropdown menus and open for having free form larger popup interfaces. This allows us to place relevant secondary information directly into the menu. This make the interface more complex, and more care needs to be asserted in designing the interfaces. However, it can also help designers negotiate screen estate better by allowing users to see necessary information when they need it instead of placing all information in the first level.

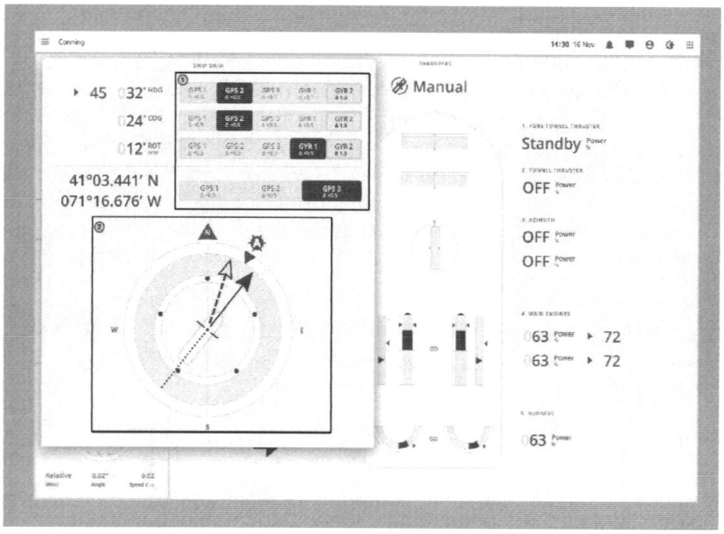

Figure 5. UI2 popup meny. Sensor settings (1), additional information (2).

4. DISCUSSION

We have demonstrated a new interaction concept for maritime interfaces that shows an example of how we might restructure user interface interaction to improve usability. Our work shows that we can significantly increase button size reduce clicks and increase information, while maintaining the information density on current displays by working on the interaction architecture of the interface itself. We have demonstrated this in the case of a conning interface. However, in our current work we have also applied the same strategy on RADAR and ECDIS interfaces, resulting in what appears to be more resilient touch interacting on these interfaces. Based on our trials, we are encouraged to move on to formal user testing of the concept in maritime operational context.

The new concept significantly alters how we structure the content in the conning interface. Although we can predict that the first step is likely to require slightly less cognitive load than selecting a particular button, the next step would arguably add more cognitive load as the user needs to reorient their attention to understand the richer information displayed in stage two. Additionally, the larger menu offers the possibility of combining more relevant information in the menu (Figure 5,2). Thus, the entire area might serve two purposes: interacting with the system and providing more information about data that cannot fit in the first-level conning interface. This adds a layer of new functionality to the conning display. However, further research is necessary to better understand how this affect users and whether it is useful in practice.

The change in information architecture is a weakness in our current understanding of the new button system. It is possible to predict that the new buttons will reduce click speed and potentially can reduce click error rates and the number of button clicks. However, since the new concept also changes the cognitive load and has a sufficiently different information architecture than the old system, there is a significant uncertainty whether this concept works in practice. More testing in real conditions is necessary to better understand the connection between information architecture, function clustering, and click surfaces.

How touch screens affect safety is a complex question that depends on many factors, such as the type of equipment, the effectiveness of existing interaction mechanisms, and design choices in each instance. However, we argue that care should be taken in design to consider whether a touch screen is a safe option in each case and ensure that implementations are well adapted to the maritime context. We argue that a versatile and adaptive approach to interaction mechanisms should be utilized in maritime settings. Given that temporary impairment is prevalent, strategies to meet such requirements should be applied. Here, multimodal interaction is an appropriate approach, making multiple interaction mechanisms available that suit different types of situational impairments [1].

5. CONCLUSION

This study demonstrates that increasing the click surface areas on maritime touch interfaces through function clustering may add benefits, particularly in challenging maritime environments. By expanding the button size and integrating related functions into larger clickable areas, we can potentially improve user accuracy, reduce interaction time, and decrease the likelihood of errors, all of which are crucial in high-stakes maritime operations.

Our work suggests that the new interface design, which moves away from the traditional 15 mm button size to larger, more user-friendly buttons, could have merit in maritime interfaces. However, it also highlights the need for further empirical testing to validate these findings in real-world conditions, considering maritime context and the cognitive load associated with the new interface structure.

Overall, our proposed design concept shows promise in creating more resilient and safer maritime interfaces. By carefully considering the unique demands of maritime environments and leveraging dynamic digital screens, we think it is room for improving touch screen usability, ensuring that they are a viable option even in the most demanding conditions.

6. ACKNOWLEDGEMENTS

This research was funded by the Norwegian Research Council through the MAROFF program, grant number 320247, OpenAR. Thanks to the OpenAR industrial partners, the OICL research group and the OpenBridge community.

7. REFERENCES

1. NORDBY K, LURÅS S. Multimodal interaction for marine workplaces used as strategy to limit effect of situational impairment in demanding maritime operations. In: *International Conference on Marine Design*, London. 2015. p. 09–02.

2. HAREIDE OS. Improving Passage Information Management for the Modern Navigator. In: *Proceedings: 19th International Association of Marine Aids to Navigation and Lighthouse Authorities Conference* (IALA 2018. Incheon, South Korea; 2018.

3. Organization IM. Convention on the International Regulations for Preventing Collisions at Sea, 1972 (COLREGs. London, UK: *International Maritime Organization*; 1977.

4. IEC 60945:2002 - Maritime navigation and radiocommunication equipment and systems - General requirements - Methods of testing and required test results [Internet]. [cited 2024 Aug 4]. Available from: https://webstore.iec.ch/en/publication/3959

5. Web Content Accessibility Guidelines (WCAG) 2.1 [Internet]. [cited 2024 Aug 4]. Available from: https://www.w3.org/TR/WCAG21/

6. OpenAR [Internet]. [cited 2024 Aug 4]. Available from: https://www.oicl.no/projects/openar

7. OpenBridge Design System 5.0 [Internet]. [cited 2023 Dec 11]. Available from: https://www.openbridge.no/

8. IEC. IEC 62288:2014 - Maritime navigation and radiocommunication equipment and systems. 2014.

9. MACKENZIE, I S., Fitts' Law as a Research and Design Tool in Human-Computer Interaction. *Human–Computer Interact.* 1992 Mar;7(1):91–139.

10. MACKENZIE, I S., BUXTON W., Extending Fitts' law to two-dimensional tasks. In: Proceedings of the SIGCHI conference on Human factors in computing systems – CHI '92 [Internet]. Monterey, California, United States: ACM Press; 1992 [cited 2024 Aug 4]. p. 219–26. Available from: http://portal.acm.org/citation.cfm?doid=142750.142794

8. AUTHORS BIOGRAPHY

Kjetil Nordby is a professor of interaction design at the Oslo School of Architecture and Design. He manages the Ocean Industries Concept Lab and the OpenBridge Design System. His work primarily focuses on designing advanced maritime workplaces, emphasizing disciplines such as interaction, graphic, and industrial design. For over a decade, he has led numerous research projects in close collaboration with leading maritime industry players.

Jon Erling Fauske is a designer and researcher in Ocean Industries Concept lab at Oslo School of Architecture and Design. His work is focusing on advanced interaction design for the maritime industries, and he is design lead on the OpenBridge Design System.

AGENT TRANSPARENCY AND HUMAN PERFORMANCE IN THE CONTEXT OF AUTONOMOUS COLLISION AVOIDANCE

G K van de Merwe, DNV & University of South-eastern Norway, Norway
S Nazir, University of South-eastern Norway, Norway
S Mallam, Memorial University of Newfoundland, Canada & University of South-eastern Norway, Norway
Ø Engelhardtsen, DNV, Norway

EXTENDED ABSTRACT

1. INTRODUCTION

In the future of maritime transport, Artificial Intelligence (AI)-enabled systems may enable ships to operate without direct human involvement [1], [2]. Given the safety-critical nature of maritime navigation, these AI-enabled systems must demonstrate a high degree of reliability and robustness across a wide range of situations. However, due to the limitations of such systems in novel and complex scenarios, careful design, implementation, management, and operation are required when deploying them in real-world environments [3], [4]. Consequently, proposed autonomous ship concepts typically involve human operators to monitor, supervise, and potentially intervene to ensure the required performance and safety levels are achieved [5], [6].

Decades of research have demonstrated significant human performance challenges associated with assigning humans a supervisory role over highly automated systems [7], [8]. When operators are tasked with supervising systems that make their own decisions and actions, they are often removed from much of the information and decision-making loop. Consequently, they may find it challenging to evaluate, understand, and predict the behavior of such systems. Additionally, passive information processing can lead to complacent behavior [9], [10], biases in decision-making [11], a reduced ability to detect critical information [12], over- or under-reliance on the system [13], and high workload when switching from supervised to manual control [14]. As a result, their capability to intervene is affected.

However, recent studies indicate that making a system's decisions, intended actions, and internal reasoning visible to the operator—that is, increasing the system's "transparency"—could help mitigate some of these issues [15]. Given the recent integration of AI-enabled systems in safety-critical domains, experience with transparency effects in these contexts is limited. Consequently, a research project was launched to investigate how transparency can aid supervisory control and how it enhances human performance in this area. This paper aims to offer a brief and concise overview of the documented research activities in [16].

2. METHODS

The studies performed in this project applied a mix of quantitative and qualitative methods. First, a Systematic Literature Review (SLR) was performed exploring experimental results from 17 peer-reviewed articles addressing the relationship between transparency and key human factors variables: Situation Awareness (SA), mental workload, and task performance [17]. Second, a Goal-Directed Task Analysis (GDTA) was performed with input from in-situ observations, -interviews onboard passenger ferries with 11 licensed navigators, and a detailed study of the collision regulations (COLREGs) [18]. Third, a model for human information processing provided the basis for a transparency model [19]. Fourth, realistic traffic situations and Human Machine Interfaces were developed based on an iterative design process and workshops with 5 navigators [20]. Finally, a controlled experiment was performed with 34 licensed navigators, assessing the effects of levels of transparency on SA, mental workload, and task performance [21], [22].

3. RESULTS

The results from the SLR showed a promising effect of transparency on SA and task performance, without affecting mental workload, for studies where participants were responding to proposals or supervising automation [17]. The study suggested that strategies to improve human performance, when interacting with intelligent agents, should focus on allowing humans to see into its information processing stages, considering the integration of information in existing HMI solutions.

The GDTA mapped and analyzed the goals, decisions, and cognitive tasks associated with conventional- and supervised collision and grounding avoidance [18]. The results provide a detailed analysis of the change in information requirements from conventional- to supervised collision avoidance. The study explored the shift in cognitive activities when the navigator's task changes from performing collision avoidance to supervising a system performing collision avoidance. To support operators in this change, explicit information requirements were identified that should allow for insight into the agent's decisions, planned actions, and underlying reasoning.

The model of information processing was used to structure the data from the GDTA and generate layers of transparency for a hypothetical CAGA system [19]. The model by Parasuraman, Sheridan, and Wickens (PSW) [23] was contextualized to the maritime collision avoidance setting such that the information from the GDTA could be structured into unique and distinct layers. It was suggested that this model may serve as a framework for transparent design as the steps in the model could represent the agent's input parameters, analysis, decisions, and planned actions. Using the model in this way, a minimum set of information requirements was made, organized in layers per processing step, resulting in a model for transparency.

Based on the structured information requirements, HMI concepts were developed for making an autonomous collision avoidance system transparent for its supervisor [20]. Here, traffic situations and symbology were developed to operationalize transparency for a hypothetical CAGA system. The symbology was integrated into the primary task display for collision avoidance, i.e., the radar display, and the PSW model was used to create distinct levels of transparency. This activity provided the groundwork for the empirical evaluation of agent transparency in a maritime collision avoidance context. In addition, the results demonstrated the value of the PSW model as a design framework for creating levels of transparency for autonomous agents.

Finally, the effect of transparency on SA, mental workload, and task performance was evaluated using a controlled experiment [21]. Based on the PSW model and the traffic situations, four levels of transparency and two levels of traffic complexity were varied in an experiment with 34 navigators. The results demonstrated a positive effect of transparency on SA without affecting mental workload. However, the time to comprehend the provided information increased with increased levels of transparency. These results indicate the benefits of applying transparency principles to autonomous collision avoidance systems in terms of SA, but care should be taken in time-critical conditions where the added transparency information may affect timely decision making. Furthermore, participants clearly indicated a preference for levels of transparency where the system's risk analysis was depicted [22]. Finally, considering the absence of the effect of transparency on mental workload, these results also indicate the value of proper HMI design through applying a structured and systematic human-centered design process [24].

4. DISCUSSION

This research project investigated the role of agent transparency in supervisory control and contributed with knowledge, methods, and tools regarding transparency in general and its application to the maritime domain specifically. As humans are foreseen to play a critical role in overseeing the functioning of AI-enabled systems, the operator's ability to understand, predict, and evaluate agent behaviour becomes a critical aspect of the human's supervisory task repertoire [25]. Consequently, it is essential that humans are informed and supported in making accurate decisions to enable timely and appropriate control when needed. Therefore, the aim of this research project was to generate and advance the knowledge on how supervisory control can be supported through agent transparency. It has contributed to this aim by demonstrating the importance of transparency in safety critical domains in terms of human performance, exploring the impact of autonomy on the operator's cognitive tasks, constructing a model for transparency, operationalising transparency for the maritime navigational context, and assessing its effects in a controlled experimental setting. Future work should address the applicability of transparency in time-constrained environments [26], and investigate the effect of dynamic forms of transparency, i.e., where the level of transparency can be adjusted according to the circumstances [27], on human performance in supervisory control. In addition, this research has made explicit the role-change that may be anticipated when introducing autonomous systems. The results have implications for scientific research and for the application of transparency as a design principle for autonomous agents.

5. CONCLUSION

This research advocates the relevance of affording human operators with insight into the reasoning of autonomous systems and establishes transparency as an important prerequisite on the path towards safe and effective human-supervisory control. With these new insights, meaningful human work may be created where the combined capabilities of human-agent teams can be optimised.

6. REFERENCES

1. ROLLS-ROYCE, "Rolls-Royce unveils a vision of the future of remote and autonomous shipping." Accessed: May 16, 2022. [Online]. Available: https://www.rolls-royce.com/media/press-releases/2016/pr-12-04-2016-rr-unveils-a-vision-of-future-of-remote-and-autonomus-shipping.aspx

2. KONGSBERG, "Kongsberg maritime and Massterly to equip and operate two zero-emission autonomous vessels for ASKO." Accessed: Nov. 18, 2020. [Online]. Available: https://www.kongsberg.com/maritime/about-us/news-and-media/news-archive/2020/zero-emission-autonomous-vessels/

3. ENDSLEY, M R., "Ironies of artificial intelligence," *Ergonomics*, vol. 0, no. 0, pp. 1–13, 2023, doi: 10.1080/00140139.2023.2243404.

4. LITTMAN, M., *et al.*, *Gathering Strength, Gathering Storms: The One Hundred Year Study on Artificial Intelligence (AI100) 2021 Study Panel Report*. Stanford, CA: Stanford University, 2021. [Online]. Available: http://ai100.stanford.edu/2021-report

5. ALSOS, O., *et al.*, "NTNU Shore Control Lab: Designing shore control centres in the age of autonomous ships," *J Phys Conf Ser*, vol. 2311, p. 012030, Jul. 2022, doi: 10.1088/1742-6596/2311/1/012030.

6. PORATHE, T., FJORTOFT, K., and BRATBERGSENGEN, I L., "Human Factors, autonomous ships and constrained coastal navigation," *IOP Conf. Ser.: Mater. Sci. Eng.*, vol. 929, no. 1, p. 012007, Nov. 2020, doi: 10.1088/1757-899X/929/1/012007.

7. ENDSLEY, M R., "From Here to Autonomy: Lessons Learned from Human-Automation Research," *Hum Factors*, vol. 59, no. 1, pp. 5–27, 2017, doi: 10.1177/0018720816681350.

8. BAINBRIDGE, L., "Ironies of automation," *Automatica*, vol. 19, no. 6, pp. 775–779, 1983, doi: 10.1016/0005-1098(83)90046-8.

9. PARASURAMAN, R., and MANZEY, D H., "Complacency and Bias in Human Use of Automation: An Attentional Integration," *Hum Factors*, vol. 52, no. 3, pp. 381–410, Jun. 2010, doi: 10.1177/0018720810376055.

10. WICKENS, C D., CLEGG, B A., VIEANE, A Z., and SEBOK, A L., "Complacency and Automation Bias in the Use of Imperfect Automation," *Hum Factors*, vol. 57, no. 5, pp. 728–739, Aug. 2015, doi: 10.1177/0018720815581940.

11. MOSIER, K., and SKITKA, L., "Human Decision Makers and Automated Decision Aids: Made for Each Other?," in *Automation and human performance: Theory and applications*, vol. 40, R. Parasuraman and M. Mouloua, Eds., Lawrence Erlbaum Associates, Inc., 1996, pp. 201–220.

12. METZGER, U., and PARASURAMAN, R., "The role of the air traffic controller in future air traffic management: An empirical study of active control versus passive monitoring," *Human Factors*, vol. 43, no. 4, pp. 519–528, 2001, doi: 10.1518/001872001775870421.

13. PARASURAMAN, R., and RILEY, V., "Humans and Automation: Use, Misuse, Disuse, Abuse," *Hum Factors*, vol. 39, no. 2, pp. 230–253, Jun. 1997, doi: 10.1518/001872097778543886.

14. ONNASCH, L., WICKENS, C D., LI, H., and MANZEY, D., "Human Performance Consequences of Stages and Levels of Automation: An Integrated Meta-Analysis," *Hum Factors*, vol. 56, no. 3, pp. 476–488, May 2014, doi: 10.1177/0018720813501549.

15. BHASKARA, A., SKINNER, M., and LOFT, S., "Agent Transparency: A Review of Current Theory and Evidence," *IEEE Trans Hum Mach Syst*, vol. 50, no. 3, pp. 215–224, Jun. 2020, doi: 10.1109/THMS.2020.2965529.

16. VAN DE MERWE, K., "Agent Transparency and Human Performance in Supervisory Control," Doctoral thesis, University of South-Eastern Norway, 2024. Accessed: Aug. 12, 2024. [Online]. Available: https://openarchive.usn.no/usn-xmlui/handle/11250/3145497

17. VAN DE MERWE, K., MALLAM, S., and NAZIR, S., "Agent Transparency, Situation Awareness, Mental Workload, and Operator Performance: A Systematic Literature Review," *Hum Factors*, vol. 66, no. 1, pp. 180–208, Jan. 2024, doi: 10.1177/00187208221077804.

18. VAN DE MERWE, K., MALLAM, S., NAZIR, S., and ENGELHARDTSEN, Ø., "Supporting human supervision in autonomous collision avoidance through agent transparency," *Safety Science*, vol. 169, p. 13, Jan. 2024, doi: 10.1016/j.ssci.2023.106329.

19. VAN DE MERWE, K., MALLAM, S., ENGELHARDTSEN, Ø, and NAZIR, S., "Towards an approach to define transparency requirements for maritime collision avoidance," *Proceedings of the Human Factors and Ergonomics Society Annual Meeting*, vol. 67, no. 1, pp. 483–488, Sep. 2023, doi: 10.1177/21695067231192862.

20. VAN DE MERWE, K., MALLAM, S., ENGELHARDTSEN, Ø, and NAZIR, S, "Operationalising Automation Transparency for Maritime Collision Avoidance," *TransNav, International Journal on Marine Navigation and Safety of Sea Transportation*, vol. 17, no. 2, Jun. 2023, doi: 10.12716/1001.17.02.09.

21. VAN DE MERWE, K., MALLAM, S., NAZIR, S., and ENGELHARDTSEN, Ø., "The Influence of Agent Transparency and Complexity on Situation Awareness, Mental Workload, and Task Performance," *Journal of Cognitive Engineering and Decision Making*, vol. 18, no. 2, pp. 156–184, Mar. 2024, doi: 10.1177/15553434241240553.

22. VAN DE MERWE, K., MALLAM, S., NAZIR, S., and ENGELHARDTSEN, Ø., "Navigators' Perspective on Information Requirements for Supervisory Control of Autonomous Ships," *Proceedings of the Human Factors and Ergonomics Society Annual Meeting*, p. 10711813241261686, Aug. 2024, doi: 10.1177/10711813241261686.

23. PARASURAMAN, R., SHERIDAN, T B., and WICKENS, C D., "A model for types and levels of human interaction with automation," *IEEE Trans Syst Man Cybern*, vol. 30, no. 3, pp. 286–297, May 2000, doi: 10.1109/3468.844354.
24. ISO, *ISO 9241-210:2019 Ergonomics of human-system interaction - Part 210: Human-centred design for interactive systems*, 2019.
25. ENDSLEY, M R., "Supporting Human-AI Teams: Transparency, explainability, and situation awareness," *Computers in Human Behavior*, vol. 140, p. 107574, Mar. 2023, doi: 10.1016/j.chb.2022.107574.
26. TATASCIORE M., and LOFT, S., "Can increased automation transparency mitigate the effects of time pressure on automation use?," *Applied Ergonomics*, vol. 114, p. 104142, Jan. 2024, doi: 10.1016/j.apergo.2023.104142.
27. VERED, M., HOWE, P., MILLER, T., SONENBERG, L., and VELLOSO, E., "Demand-Driven Transparency for Monitoring Intelligent Agents," *IEEE Transactions on Human-Machine Systems*, vol. 50, no. 3, pp. 264–275, Jun. 2020, doi: 10.1109/THMS.2020.2988859.

7. AUTHORS BIOGRAPHY

G K van de Merwe is a principal researcher at DNV Group R&D at Høvik, Norway. He received his MSc in Cognitive Psychology in 2004, an MSc in Industrial Ecology in 2006 from Leiden University, The Netherlands. He received his PhD in Nautical Operations from the University of South-eastern Norway in 2024.

S Nazir is a Professor in Training and Assessment at Department of Maritime Operations at the University of South-Eastern Norway. He received his PhD in Industrial Chemistry and Chemical Engineering from Politecnico di Milano, Italy, in 2014.

S Mallam is an Associate Professor of Maritime Human Factors at the Fisheries and Marine Institute, Memorial University of Newfoundland, Canada, and at the Faculty of Technology Natural Sciences and Maritime Sciences at The University of South-Eastern Norway. He received his PhD in Human Factors in 2016 from Chalmers University of Technology, Sweden.

Ø Engelhardtsen is group lead of the Ship Autonomy section at DNV Group R&D at Høvik, Norway. He is trained as a commanding officer in the Royal Norwegian Navy. He received his MSc. in Cybernetics at the Norwegian University of Science and Technology in 2007.

Human Factors 2024
8th – 9th October 2024
Wageningen, Netherlands

OPPORTUNITIES FOR ADVANCED MAN-MACHINE TEAMING ON SHIPS THROUGH HOLONIC HUMAN CYBER-PHYSICAL SYSTEMS

NC Taylor, Department of Mechanical and Mechatronic Engineering, Stellenbosch University, South Africa
K Kruger, Institute for Manufacturing, University of Cambridge, UK
A Bekker, Department of Mechanical and Mechatronic Engineering, Stellenbosch University, South Africa

SUMMARY

Teamwork among humans on a ship is essential for performing many tasks aboard. Working towards the successful completion of vessel operations regularly requires humans to interact with interfaces to digital systems and machines, forming man-machine teams. This article explores the potential opportunities for advanced man-machine teaming on ships that can be unlocked through the development of holonic human cyber-physical systems. Human cyber-physical systems are state-of-the-art information and communications technology wherein physical assets and humans are equipped with digital counterparts. A holonic systems approach is used whereby autonomous, cooperative elements, called holons, are created to represent entities that are in the real world, including humans. An illustrative example is presented to exemplify an implementation of a holonic human cyber-physical system towards creating a decision support system with human condition monitoring.

KEYWORDS

Man-machine teaming; Shipping; Human cyber-physical systems; Holonic systems

1. INTRODUCTION

Human factors engineering is seen as an integral branch of human factors, which focusses on the integration of human characteristics, capabilities, skills and limitations, as well as the job they are required to do, into the definition, design and development of systems [1]. This is realised through:

- adoption of a systems approach; and
- focus on optimisation of human performance and well-being, which are closely related [2].

It is evident that vessel designers have considered human factors during ship design, for example by making ship accommodation more comfortable and adhering to safety measures [3]. However, traditional ship design practices are criticised for their mechanistic approaches, insufficiently accounting for the human element [4].

In a ship context, the human element is comprised of that which influences the interaction between a human and another entity on a vessel, which can be another human, system or machine [1]. The human element as a user has not been forgotten in ship design practices. Many human-centred systems have been defined in principle and as approaches, such as human-in-the-loop systems, human-machine systems, human-machine and human-automation symbiosis, and human-machine cooperation [5-10]. Similarities are seen in the development of cyber-physical systems with human elements that include humans in control processes and complex systems, keep human decision making in the loop at different levels and enable mutual communication (not only providing information to a user). However, in ship digitalisation progress, humans remain constrained to the physical layer and are not integrated into the digital environment. Humans do not have digital counterparts that serve as digital representatives for them in largely digitalised ship contexts. Fortunately, ship design approaches appear to be shifting from technically centred to human-centred approaches, but persistent efforts are still required to get the human-centred rationale and design to permeate into practice in existing technically biased systems and processes [11].

Human cyber-physical systems could provide a means for integrating human factors into a digital environment [12]. Following a human factors engineering approach, systems engineering design is recommended for the development of human cyber-physical systems for seafarers and is focussed on enhancing task performance shipboard through managing the condition or well-being of seafarers through human digital twins. The development of a human cyber-physical system on a ship necessitates equipping seafarers with digital counterparts (human digital twins) that uniquely represent each seafarer's state and behaviour, facilitates interaction with their digital counterparts and supports communication with other digital systems through their digital counterparts. Creating human digital twins for seafarers can provide an effective platform for human-centred computing towards seafarer motion sickness monitoring in a human cyber-physical system [13].

This work explores a similar premise for the integration of seafarers into a digital environment through a holonic systems approach. Holonic systems are constructed with building blocks, called holons, that are autonomous and cooperative elements comprising of information processing parts and, optionally, physical parts [14]. The physical parts can include machines and humans, which the information processing parts of a holonic system are intended to digitally represent and assist towards achieving common goals.

Man-machine teaming involves members of a group working together towards achieving a common goal, where the group members are comprised of humans and other agents, including autonomous systems and machines [15]. Work aboard ships exists in a team structure with hierarchical management, which research shows is an effective manner for humans to organise themselves and perform tasks. Several options exist for structuring members of a group for man-machine teaming, such as division of labour through task allocation, but the best way is still a subject of relevant study [15].

This article presents opportunities for the incorporation of humans in the development of holonic human cyber-physical systems that are aimed toward advanced man-machine teaming on ships. The article starts by examining the current state of man-machine teaming on ships, then holonic human cyber-physical systems concepts as a foundation for exploring the potential they may offer for man-machine teaming in a ship context. Thereafter, the opportunities that holonic human cyber-physical systems may present for advanced man-machine teaming on ships are identified. An example of a holonic human cyber-physical system in action on board a ship for a man-machine teaming activity is then presented. Lastly, a discussion is offered centred around the implications that identified opportunities suggest for man-machine teaming on a ship, before the conclusions are drawn.

2. MAN-MACHINE TEAMING ON SHIPS

The ship context is under transformation due to technological developments; today, vessels are equipped with sophisticated systems and multiple levels of automation [16-20]. Automation, as used in this article, refers to manual tasks previously performed by seafarers that have been automated, which differs from autonomous systems, which are self-directed and can learn without the necessary human intervention [15]. Although automation shipboard has enabled reductions in numbers required on board, human intervention is still crucial for making decisions and controlling actions during many vessel operations [21, 22]. Using automation to enhance seafarer task performance, instead of pursuing fully autonomous and unmanned vessel operations, seems to be a preferred trend [23].

Shipboard automation efforts are resulting in the workload of seafarers increasingly moving from physical to cognitive tasks. Seafarers are expected to coordinate several sources of information and understand the vessel and its environment more through man-machine interfaces [11, 24]. In this sense, automation is used as a tool, but man-machine teaming, where systems with different levels of autonomy act as members of the seafaring team, is not truly realised. In fact, research shows that an increased demand to integrate information from multiple sources, where automation is used as a supportive aid, can amplify the cognitive load on human operators, rather than reduce their workload [15].

Interactions between humans and systems with varying levels of autonomy can be described through patterns, such as structures, modes and forms [25]. These patterns describe who (or what) is in control of a task that needs to be assigned and executed, based on the autonomy, ability, authority and responsibility of the participating human and system. The patterns enable man-machine teaming at different levels of autonomy, operating in the space between tasks that are completely manual (performed by man) or fully autonomous (performed by a system, which can be comprised of automated machines).

3. HOLONIC HUMAN CYBER-PHYSICAL SYSTEMS

Human cyber-physical systems are considered state-of-the-art information and communications technology that incorporate human factors into the development and integration of systems comprising of cyber and physical layers in which humans play integral roles [26-29]. Cyber layers can consist of software computations and decision making; physical layers have physical infrastructures and processes, where humans are physically present [28, 29]. Human cyber-physical systems are typically designed for applications in areas of infrastructure development where the well-being of people is an important consideration, such as smart cities and healthcare [29].

Human cyber-physical systems are aimed at fostering human competencies and augmenting human abilities through technology to improve man-machine interaction [26, 27, 29]. However, human cyber-physical system advances do not provide sufficient guidance for system organisation, which is necessary for forming and utilising man-machine teams. The concept of holonic systems, which originates from the work of Koestler [30], offers a means for the dynamic organisation of human cyber-physical systems.

Koestler introduced the term *holon* that derives from the Greek word *holos* and suffix *-on*, meaning whole and part, respectively [30]. Therefore, the holon conceptually reflects the dual nature of an intelligent element that is a self-contained whole in itself and a dependent part of a greater self-organising system, such as cells in biological systems, which Koestler drew inspiration from. A holon consists of an information processing part and can contain a physical processing part, be composed of other holons or be part of another holon [14]. Practically, a holon is a building block of a holonic system for transforming, transporting, storing and validating information and physical assets [14].

A self-organising, hierarchical system of holons is defined as a *holarchy* [30]. In functioning independently, yet cooperatively, and being part of another holon or a holarchy, holons are characterised by *autonomy*, *cooperation* and *decomposability* [14, 30]. Autonomy is the ability of holons to control themselves and execute plans or strategies that they create for themselves. Cooperation is the capability of a set of holons to develop and execute mutually acceptable plans or strategies. Decomposability refers to the construction of complex holons that comprise of constituent holons, implying that elementary level holons exist as foundational building blocks for other holons. These properties of holons enable the formation of holarchies that can be tailored to dynamic environments for various purposes that contexts may require. Holonic systems have proven to be effective in the manufacturing, healthcare, mining and agriculture domains [31-37]

Human resource holons are proposed to integrate humans in a holonic system [34, 37]. In a human resource holon, a human is the physical part associated with an information processing part, and possibly human interfaces, dedicated to an individual person. Interface holons that are devoted to providing human interfaces for multiple humans, i.e. not allocated to a single human such as a wearable sensor might be, can additionally be used to facilitate communication between multiple humans and their coupled information processing parts in the holonic system.

4. OPPORTUNITIES FOR ADVANCED MAN-MACHINE TEAMING ON SHIPS

Opportunities that holonic systems may provide towards man-machine teaming on a ship are generated on the premise that each machine *and* human on a ship is to be equipped with an information processing part to form resource holons. It is imperative that human resource holons are formed alongside other (non-human) resource holons – human resource holons serve as the cornerstone for integrating human factors during the development of a holarchy in a ship context.

4.1 HUMAN FACTOR INTEGRATION

The information processing parts of holons purposely represent the world of interest they are part of, enabling the reflection of the world of interest in the holonic human cyber-physical system. Each human resource holon is responsible for digitally representing the state and behaviour of their associated human through the information processing part. In this way, human factors are integrated into the holonic human cyber-physical system for a ship at the design phase of the holarchy, which dynamically evolves with humans as time progresses through the operation phase of the holarchy.

The human resource holons can additionally be personalised with data pertaining to their human's mandated work tasks to be performed and their position on board, as well as their status of health, comfort and well-being, such as motion sickness or fatigue levels. In this way, the human resource holon can serve as a human-centred health monitoring system for humans on board. The perception of the holonic system from the human's perspective and what data the holonic system keeps track of during operation are both important factors to consider, as they, respectively, influence trust in the system and data privacy.

4.2 MAN-MACHINE COOPERATION

Levels of man-machine cooperation have been defined [38]. The levels of cooperation refer to the *know-how* of a team member (human or machine), also known as their ability to control a process, and the *know-how-to-cooperate*, which is concerned with the ability of the team member to work with others involved in the process. Interface holons and human resource holons can provide personalised interfaces for the interaction between the information processing part of a human and the human. More importantly, human resource holons ensure that other resource holons on a ship can identify the human presence on board in software systems (through the information processing counterparts of humans) as much as their physical human presence (through human interfaces). The human resource holons manage information about their associated humans such that human autonomy, ability, authority and responsibility can be accessible to resource holons of machines for task allocation and implementation. In this way, man-machine teaming can be facilitated at multiple levels as required by the context on board a ship and customised to human and machine roles, allowing for the alignment of human and machine capabilities.

4.3 DISTRIBUTION AND DELEGATION OF DECISION MAKING

Data or information processing can be allocated to machine and human resource holons according to human-machine interaction patterns that match the requirements for seafarer roles on a ship. A holonic human cyber-physical system enables the separation of concerns among holons that can be set up to reduce cognitive load and mental fatigue in humans. The information processing parts of (human and non-human) resource holons could be programmed to automate redundant and computationally laden data processing or integration tasks to allow humans to focus their attention on more pressing matters or areas that require human assistance, thereby distributing decision making among man and machine. Additionally, a holonic human cyber-physical system can support man or machine in delegating decision making to other holons that may be better suited to make the decisions based on decision-making context and the autonomy, ability, authority and responsibility best suited to make the decisions, e.g. if a particular holon is known to be more capable or knowledgeable.

4.4 TRACEABILITY OF MAN-MACHINE INTERACTIONS

Each holon can be used to keep track of interactions between their physical resource and other resources. In this way, a history of human-machine interactions can be available to improve traceability that informs inspections or audits when erroneous behaviours or accidents are investigated. The tracked interactions would help improve situational awareness for scenarios being interrogated. Traceability like this could be particularly informative when cases of human errors arise and humans are under examination despite following protocol as required to.

4.5 ADAPTABILITY TO CHANGES IN TEAMS AND ROLES

It is not uncommon that humans on board change between voyages and missions. Humans may also switch between roles or tasks on board, referred to as context switching [34]. The human resource holons can serve as profiles for humans that are allocated to their vessel (and holarchy) of operation for specific voyages. Holonic human cyber-physical systems are adaptive in nature and can handle changes in machines and humans, or their roles, without stopping operation when individual holons are added, amended or removed from the holarchy. In this way, human resource holons can be switched between holarchies on different vessels, provided each vessel in the fleet is augmented with its own holarchy, and man-machine teams on board individual vessels to aid resource management.

4.6 REACTIVITY TO CHANGES IN TASK PERFORMANCE

A holonic human cyber-physical system can increase situational and spatial awareness for both humans and machines through being perceptive of the physical environment it monitors, much like ambient intelligence environments comprised of intelligent interfaces [39]. If any changes in task performance or allocation occur, a holonic human cyber-physical system can respond accordingly without necessarily stopping overall operations. The connected basis of the holarchy supports the communication of changes to resources that need to be informed to handle amendments in plans.

5. ILLUSTRATIVE EXAMPLE

Ship navigation is used as a representative activity of man-machine teaming between an officer and a navigation system on board a vessel. The goal of moving the ship from point A (starting point) to point B (destination) along a defined route can include planning the route, monitoring the traverse along the route and making decisions based on information acquired en route until completion.

The architecture of a traditional means of man-machine interaction for ship navigation is depicted in Figure 1 to serve as a backdrop for the holonic human cyber-physical system architecture, illustrated in Figure 2. In Figure 1, the officer interacts with the navigation system (comprised of an information processing part) through an interface. The interaction is limited to inputs from the officer reaching the navigation system and the officer receiving feedback from the navigation system via the interface directly. The distinction from Figure 1 is the inclusion of an information processing part for the officer in the cyber layer in Figure 2. The officer works with the navigation system *through* their information processing part, which they interact with via an interface.

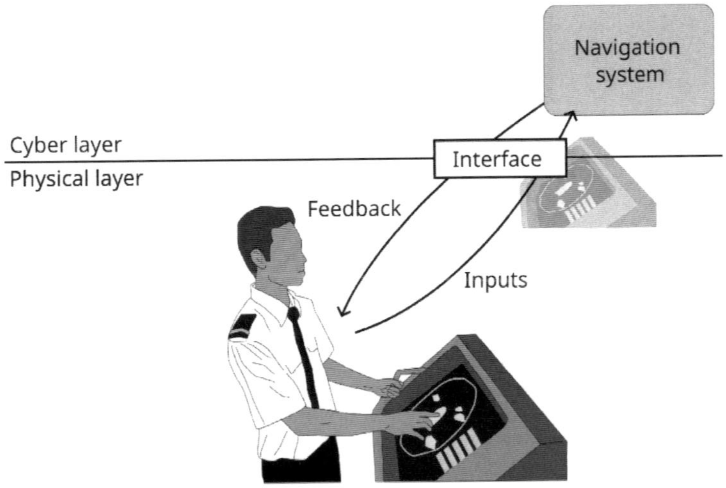

Figure 1. Traditional man-machine interaction for ship navigation.

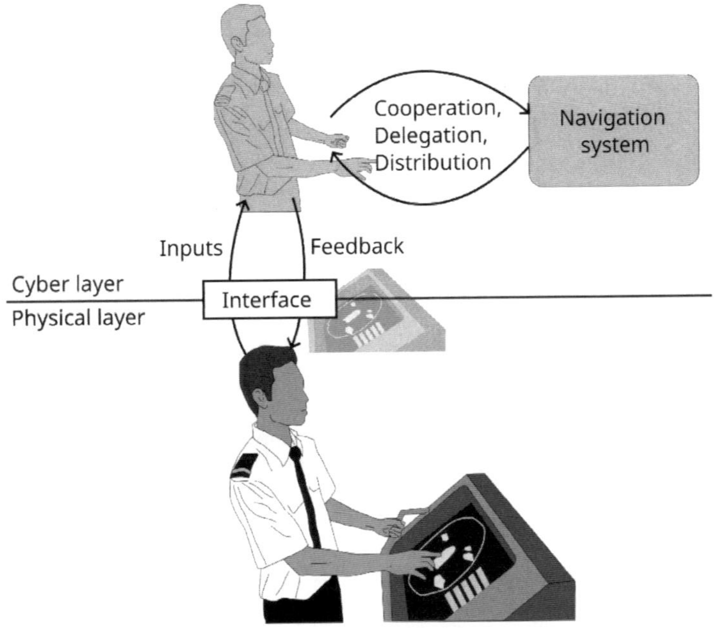

Figure 2. Man-machine teaming through a holonic human cyber-physical system for ship navigation.

The inclusion of an information processing part for the officer integrates human factors unique to the officer into the cyber layer for advanced man-machine teaming. Interaction between the officer and their information processing part can be personalised considering their respective preferences and capabilities. The information processing part will have knowledge about the autonomy, ability, authority and responsibility of the officer and enable cooperation with the navigation system. In this way, tasks can be distributed and delegated by the officer to their information processing part or the navigation system, and vice versa. For example, the officer can provide points A and B (starting point and destination for navigation), the navigation system can then compute route options, and the officer's information processing part can select the most suitable route based on preferences set by the officer.

It is important to understand from Figure 2 that the information processing part of the officer is not only facilitating cooperation. The information processing part is also reflecting the state and behaviour of the officer. Something that is not depicted in Figure 2 is the presence and interaction among the information processing parts of other humans on board and the navigation system and officer's information processing part. As each human is equipped with an information processing part on board, the navigation system and officer can be presented with the condition of all humans on board during navigation. The navigation system and officer can thus be informed if humans are becoming severely motion sick, tasks are being interrupted due to motion or health and safety thresholds are exceeded owing to vibration levels, which are typically not available digitally during operation today [40]. Moreover, the communication among the humans on board with their information processing parts and the navigation system generates channels of data and information flow enabling enhanced interaction traceability.

6. DISCUSSION

Holonic systems principles have been investigated in the manufacturing, healthcare and agriculture domains, which this article translates to a ship context. A holonic human cyber-physical system approach is applied to create holons for machines and humans shipboard that can cooperate with each other both in the physical and cyber layers. Through this process, opportunities are identified for achieving advanced man-machine teaming on ships.

The construction of human resource holons positions humans as integral components of the greater holonic human cyber-physical system, not only as users constrained to the physical layer, by providing them with personalised information processing parts. In this way, a human-centred approach is inherent to the development of the ship holarchy. Human factors are integrated into the information processing parts for each human and, thereby, made accessible to the information parts of other holons for the purposes of cooperation. Moreover, the information processing parts of humans can monitor the condition of their associated humans, implying that the holonic human cyber-physical system serves as a health monitoring platform for the conditions of humans during vessel operations, above and beyond typical shipboard health monitoring systems that keep track of the status of onboard machinery only. The illustrative example demonstrates this through a practical scenario where human factors can be quantifiably considered during navigational decision making.

Effective man-machine teaming requires the representation of the autonomy, ability, authority and responsibility of each cooperating entity. These attributes can be represented within the information processing parts of human and non-human resources. Equipping humans with their own information processing parts, in addition to machines, lends itself to facilitating the distribution and delegation of decision making among humans and machines, as well as between physical and information processing parts of resources. To this end, cooperation between physical parts (whether machine or human) and associated information processing parts, as well as inter-holon cooperation, can be assisted through a holonic human cyber-physical system.

A human cyber-physical system may also enhance interactions through personalised man-machine interfaces, where interface holons play a crucial role, and improve the traceability of man-machine interactions. The rationale for enriching man-machine interactions and traceability of interactions on board includes fostering areas where human errors are historically prone to augment human task performance, not intentionally form evidence for blaming workers when accidents happen. In light of this, and in addition to autonomy, ability, authority and responsibility that man-machine teaming patterns may incorporate, ethics and trust are important factors to consider during implementation [10]. Other concerns raised could include shifts in traditional seafarer work responsibilities, potential overreliance on machines and a resultant reduction in the experience or understanding gained by humans during situations that would previously demand their direct attention and problem-solving skills.

The adaptive and reactive nature of a holonic human cyber-physical system is supportive of a dynamic working and living environment that is typically present on board a ship. The holonic human cyber-physical system can respond actively to accommodate changes in tasks and schedules, but also resources allocated to tasks as required. Much of this article has focussed on reflecting and supporting operations in real time, but the holonic human cyber-physical system could be used to run simulations of future tasks to provide foresight into decisions. For example, voyage routing options can be weighed up along multiple preferred routes considering the resulting human conditions and anticipated task performance influenced by ship responses according to weather forecasts. This simulation functionality can be extended further, creating and configuring holonic ship and human cyber-physical systems with hypothetical physical parts (including ship structures, machines fitted shipboard and expected cohorts of humans) during the ship design stage to assess ship seakeeping performance and human responses well in advance.

Amidst the opportunities identified, a notable concern around complexity of implementation should be highlighted. There are multiple interacting resources on board a ship that can be complicated to map accurately through information processing parts of holons. Although holonic human cyber-physical systems are by nature scalable, comprehensive domain knowledge is required alongside large system development and configuration efforts to realise an appropriately functional holonic human cyber-physical system. The reflection of activities in the physical environment on board a ship is by no means a trivial task to mirror in a virtual realm. Significant development time will be required to realise an effective holonic human cyber-physical system, which translates into the affinity for adoption of the development and use of holonic human cyber-physical systems during ship processes that are already well defined. A potential solution to the implementation complexity is to consider federations of holons for humans or machines with similar roles on board, such as grouping all passengers, negating the detailed implementation for every single resource on board.

7. CONCLUSIONS

Holonic systems are arrangements of foundational building blocks, known as holons, that have an information processing part and likely a physical part. In this article, opportunities that holonic human cyber-physical systems present for advanced man-machine teaming on a ship were identified and discussed. The opportunities highlight that autonomy and cooperation characteristics of holons appear to be well aligned with enabling resource task allocation and performance that leads towards advanced man-machine teaming in a ship context. The integration of human resource holons for humans in the development of a holonic human cyber-physical system for a ship denotes a human-centred approach, positioning humans as integral elements of a ship holarchy and not merely as users of another cyber-physical system installed shipboard.

8. REFERENCES

1. Lloyd's Register, 'The Human Element – An Introduction', *The Lloyd's Register Group*, 2008.

2. International Ergonomics Association and International Labour Organisation, 'Principles and Guidelines for Human Factors/Ergonomics Design and Management of Work Systems', *International Ergonomics Association and International Labour Organisation*, 2021.

3. RUMAWAS, V., 'Human Factors in Ship Design and Operation: Experiential Learning', *Doctoral Dissertation at Norwegian University of Science and Technology*, 2016.

4. ABEYSIRIWARDHANE, A., 'Shaping Ships for People: Human Centred Design Knowledge into Maritime Education', *Doctoral Dissertation at University of Tasmania*, 2017.

5. TZAFESTAS, S.G., 'Concerning Human-Automation Symbiosis in the Society and the Nature', *International Journal of Factory Automation, Robotics and Soft Computing*, (3):16-24., 2006.

6. SCHIRNER, G., ERDOGMUS, D., CHOWDHURY, K., PADIR, T., 'The Future of Human-in-the-Loop Cyber-Physical Systems', *Computer*, 46(1):36-45, 2013.

7. PACAUX-LEMOINE, M-P., TRENTESAUX, D., REY, G.Z., MILLOT, P., 'Designing Intelligent Manufacturing Systems through Human-Machine Cooperation Principles: A Human-Centred Approach', *Computers and Industrial Engineering*, 111:581-595. 2017.

8. JIRGL, M., BRADAC, Z., FIEDLER, P., 'Human-in-the-Loop Issue in Context of the Cyber-Physical Systems', *International Federation of Automatic Control PapersOnLine*, 51(6):225-230, 2018.

9. PACAUX-LEMOINE, M-P., BERDAL, Q., ENJALBERT, S., TRENTESAUX, D., 'Towards Human-Based Industrial Cyber-Physical Systems', *Proceedings of the Institute of Electrical and Electronics Engineers Industrial Cyber-Physical Systems*, pp. 615-620, 2018.

10. PACAUX-LEMOINE, M-P., TRENTESAUX, D., 'Ethical Risks of Human-Machine Symbiosis in Industry 4.0: Insights from the Human-Machine Cooperation Approach', *Proceedings of the 14th International Federation of Automatic Control Symposium on Analysis, Design and Evaluation of Human Machine Systems*, pp. 19-24, 2019.

11. VU, V.D., LÜTZHÖFT, M.H., 'Improving Human-Centred Design Application in The Maritime Industry – Challenges and Opportunities', *Proceedings of the Royal Institution of Naval Architects Human Factors Conference*, 2020.

12. TAYLOR, N.C., BEKKER, A., KRUGER, K., 'Mariner 4.0: Integrating Seafarers into a Maritime 4.0 Environment', *International Journal of Maritime Engineering*, 164(A4):373-384, 2023.

13. TAYLOR, N.C., KRUGER, K., BEKKER, A., 'A Human Cyber-Physical System for Human-Centred Computing in Seafaring', *Journal of Ambient Intelligence and Humanized Computing*, 14:7871-7884, 2023.

14. CHRISTENSEN, J.H., 'Holonic Manufacturing Systems: Initial Architecture and Standards Directions', *First European Conference on Holonic Manufacturing Systems*, Hannover, Germany, 1994.

15. WALLISER. J.C., DE VISSER, E.J., WIESE, E., SHAW, T.H., 'Team Structure and Team Building Improve Human-Machine Teaming with Autonomous Agents', *Journal of Cognitive Engineering and Decision Making*, 13(4):258-278, 2019.

16. FELSKI, A., ZWOLAK, K., 'The Ocean-Going Autonomous Ship – Challenges and Threats', *Journal of Marine Science and Engineering*, 8(1), 41, 2020.

17. Lloyd's Register, 'Cyber-Enabled Ships: ShipRight Procedure – Autonomous Ships', *Lloyd's Register*, 2016.

18. DNV GL, 'Class Guideline: Autonomous and Remotely Operated Ships', *DNVGL-CG-0264*, 2018.

19. International Maritime Organisation, 'Outcome of the Regulatory Scoping Exercise for the Use of Maritime Autonomous Surface Ships (MASS)', *International Maritime Organisation*, 2021.

20. Bureau Veritas, 'Guidelines for Autonomous Shipping', *Bureau Veritas*, 2019.

21. KURT, R.E., KHALID, H., TURAN, O., HOUBEN, M., BOS, J., HELVACIOGLU, I.H., 'Towards Human-Oriented Norms: Considering the Effects of Noise Exposure on Board Ships', *Ocean Engineering*, 120:101-107, 2016.

22. Lloyd's Register. 'The Human-Centred Approach – A Best Practice Guide for Equipment Manufacturers', *The Lloyd's Register Group*, 2014.

23. LÜTZHÖFT, M., HYNNEKLEIV, A., EARTHY, J.V., PETERSEN, E.S., 'Human-Centred Maritime Autonomy – An Ethnography of the Future, *Journal of Physics: Conference Series*, 1357, 2019.

24. Lloyd's Register. 'The Human-Centred Approach – A Best Practice Guide for Ship Designers', *The Lloyd's Register Group*, 2014.

25. PACAUX-LEMOINE, M-P., FLEMISCH, F., 'Layers of Shared and Cooperative Control, Assistance and Automation', *Cognition, Technology and Work*, 21:579-591, 2019.

26. ROMERO, D., BERNUS, P., NORAN, O., STAHRE, J., FAST-BERGLUND, A., 'The Operator 4.0: Human Cyber-Physical Systems and Adaptive Automation Towards Human-Automation Symbiosis Work Systems', *Proceedings of the International Federation for Information Processing on Advances in Production Management Systems*, pp. 677-686, 2016.

27. ROMERO, D., STAHRE, J., WUEST, T., NORAN, O., BERNUS, P., FAST-BERGLUND, A., GORECKY, D., 'Towards an Operator 4.0 Typology: A Human-Centric Perspective on the Fourth Industrial Revolution Technologies', *Proceedings of the International Conference on Computers and Industrial Engineering*, pp. 1-11, 2016.

28. PUTNIK, G.D., FERREIRA, L., LOPES, N., PUTNIK, Z., 'What is a Cyber-Physical System: Definitions and Models Spectrum', *Faculty of Mechanical Engineering Transactions*, 47:663-674, 2019.

29. LIU, Z., WANG, J., 'Human-Cyber-Physical Systems: Concepts, Challenges and Research Opportunities', *Frontiers of Information Technology and Electronic Engineering*, 21(11):1535-1553, 2020.

30. KOESTLER, A., 'The Ghost in the Machine', *The Macmillan Company*, 1967.

31. DEFTY, T., KRUGER, K., BASSON, A., 'A Design Process for Holonic Systems Using the ARTI and BASE Architectures', In: Borangiu, T., Trentesaux, D., Leitão, P., Berrah, L., Jimenez, J-F. (eds.), Service Oriented, Holonic and Multi-agent Manufacturing Systems for Industry of the Future, SOHOMA 2023, *Studies in Computational Intelligence*, 1136:62-73, Cham: Springer.

32. ERWEE, H., BASSON, A., KRUGER, K., 'A Holonic Approach to Clinical Pathway Data Analysis', *Computers in Biology and Medicine*, 181:109073, 2024.

33. VAN NIEKERK, D., KRUGER, K., BASSON, A., 'Interface Holons in the BASE Architecture for Human-System Integration in Cyber-Physical Systems', *Proceedings of the 8th International Conference on Competitive Manufacturing*, 2022.

34. SPARROW, D.E., KRUGER, K., BASSON, A., 'An architecture to Facilitate the Integration of Human Workers in Industry 4.0 Environments', *International Journal of Production Research*, 2021.

35. CADER, A., KRUGER, K., BASSON, A.H., 'Holonic Systems for Traceability in Fruit Supply Chains', In: Borangiu, T., Trentesaux, D., Leitão, P., Berrah, L., Jimenez, J-F. (eds.), Service Oriented, Holonic and Multi-agent Manufacturing Systems for Industry of the Future, SOHOMA 2023, *Studies in Computational Intelligence*, 1136:62-73, Cham: Springer.

36. DOUBELL, C.D., BASSON, A.H., KRUGER, K., CONRADIE, P.D.F., 'A Digital Twin System for Railway Infrastructure', *R&D Journal*, 39:23-34, 2023.

37. LEUVENNINK, J.C., 'Architectures for Human Worker Integration in Holonic Manufacturing Systems', In: Borangiu, T., Trentesaux, D., Thomas, A., Cavalieri, S. (eds.), Service Oriented, Holonic and Multi-agent Manufacturing Systems for Industry of the Future, SOHOMA 2018, *Studies in Computational Intelligence*, 803:133-144, Cham: Springer.

38. PACAUX-LEMOINE, M-P., VANDERHAEGEN, F., 'Towards Levels of Cooperation', *Proceedings of the IEEE International Conference on Systems, Man, and Cybernetics*, 2013.

39. SPARROW, D.E., TAYLOR, N.C., KRUGER, K., BASSON, A., BEKKER, A., 'Interfacing with Humans in Factories of the Future: Holonic Interface Services for Ambient Intelligence Environments', In: Borangiu, T., Trentesaux, D., Leitão, P., Cardin, O., Lamouri, S. (eds.), Service Oriented, Holonic and Multi-agent Manufacturing Systems for Industry of the Future, SOHOMA 2020, *Studies in Computational Intelligence*, 952:299-312, Cham: Springer.

40. TAYLOR, N.C., BEKKER, A., KRUGER, K., 'The Operational Development of Diagnostic Seasickness Criteria through a Human Cyber-Physical System', *Applied Ergonomics*, 119:104316, 2024.

9. AUTHORS BIOGRAPHY

Dr Nicole Catherine Taylor holds the position of Lecturer at the Department of Mechanical and Mechatronic Engineering, Stellenbosch University. She is dually responsible for teaching Mechatronic Engineering students and conducting research. Her research interest aligns with the integration of human factors in advances of digital technologies, particularly concerning the development of human cyber-physical systems for seafarers.

Prof Karel Kruger holds the position of Associate Professor at the Institute for Manufacturing, University of Cambridge. His research focusses on two themes: digital twins for complex systems and human-system integration in operational environments. He has explored these themes in different application domains, including manufacturing, mining, healthcare, agriculture and maritime.

Prof Anriëtte Bekker holds the position of Professor at the Department of Mechanical and Mechatronic Engineering, Stellenbosch University. She is the director of the Sound and Vibration Research Group, a postgraduate level research group, and teaches Mechanical and Mechatronic Engineering students. Her research interests include human responses to sound and vibration in the occupational environment and vibration-related responses of polar supply and research vessels in the Arctic and Antarctic, operational modal analysis, rotational dynamics and digital twin technology.

INTERORGANIZATIONAL SIMULATOR TRAINING FOR AERONAUTICAL AND MARITIME SEARCH AND RESCUE (SAR) PERSONNEL

O Schliebusch-Jacob, German Maritime Search and Rescue Service, Germany; **T Lübcke,** German Maritime Search and Rescue Service, Germany

SUMMARY

Seamless cooperation between air and sea units is essential to deal effectively with major maritime emergencies and requires regular, realistic training, which in practice can be costly and logistically challenging.

Simulation-based training for aeronautical and maritime SAR personnel was previously only possible in isolated simulation environments. The AMARIS (Aeronautical and Maritime Innovation Environment for Interorganizational Simulation) project addresses this shortcoming.

Linking the Air Vehicle Simulator of the German Aerospace Center with the SAR simulator of the German Maritime Search and Rescue Service, AMARIS allows ships and helicopters to interact in a shared simulation world. This is the first time simulation environments have been linked across organisational boundaries and different domains.

From a human factors perspective this setup is complemented by a research-based training concept to improve cooperation and communication between ship and helicopter units. Initial research findings will be presented and practical implications for similar simulator collaborations will be discussed.

NOMENCLATURE

3D	3 dimensional
ARCC	Aeronautical Rescue Coordination Centre
AMARIS	Aeronautical and Maritime Innovation Environment for Interorganizational Simulation)
AVES	Air Vehicle Simulator
BRR	Bremen Rescue Radio
DGzRS	Deutsche Gesellschaft zur Rettung Schiffbrüchiger (German Maritime Search and Rescue Service)
DLR	Deutsches Zentrum für Luft- und Raumfahrt (German Aerospace Center)
DIS	Distributed Interactive Simulation
EMSN	European Maritime Simulator Network
EU	European Union
HEMS	Helicopter Emergency Medical Service
HLA	High Level Architecture
HRV	Heart Rate Variability
IAMSAR	International Aeronautical and Maritime Search and Rescue Manual
ICAO	International Civil Aviation Organization
IEEE	Institute of Electrical and Electronics Engineers
IMO	International Maritime Organization
LIVEX	Live Exercise
MARCOM	Allied Maritime Command
MONALISA	Motorways & Electronic Navigation by Intelligence at Sea
MOS	Mobile Operation Station
MSCW	Maritime Simulation Centre Warnemünde
MRCC	Maritime Rescue Coordination Center
NATO	North Atlantic Treaty Organization
NH90	NATO-Helicopter 90
OSC	On scene co-ordinator
SAR	Search and Rescue
SAREX	Search and Rescue Exercise
SIMNET	Simulator Networking
UW/H	Universität Witten/Herdecke
VHF	Very High Frequency
VPN	Virtual Private Network

1. INTRODUCTION

Dealing with major emergencies at sea poses a particular challenge for the search and rescue services involved. Especially when passengers and crew members are rescued from a ship by helicopter in adverse weather conditions, the coordination of the rescue teams at sea and in the air must be ensured. A modern simulation environment is required to train such frequently complex operational situations. The aim of the AMARIS project is to create a simulation environment for training complex operational situations in which flying and seagoing personnel can train in a joint training environment. While training has generally taken place in separate simulators up to now, the AMARIS project will create an environment in which flying and seagoing personnel can train together. For this purpose, the SAR simulator centre of the DGzRS (Deutsche Gesellschaft zur Rettung Schiffbrüchiger - German Maritime Search and Rescue Service) in Bremen/Germany was coupled Air Vehicle Simulator (AVES) of the DLR (Deutsches Zentrum für Luft- und Raumfahrt - German Aerospace Center) in Braunschweig/Germany and then used in networked operation. In this paper we will address both technical and human factor aspects of the innovative AMARIS training environment.

After a brief presentation of the operational background of the AMARIS research project (section 2), the use, possibilities and limitations of maritime SAR simulator training are explored. After explaining the need for creating simulator networks for maritime simulator training, existing solutions are categorized, and the innovative aspects of the AMARIS setup are analysed by outlining the technical challenges as well as the expected training effects of coupling the two simulators (section 3). Section 4 roughly sketches the procedure for developing the AMARIS training concept and presents the AMARIS training process. In section 5 preliminary results and experiences from both a technical and a behavioural science perspective, in particular with regard to the subjectively assessed training success [1] are presented. Section 6 draws initial conclusions and implications for practice based on the results. A special focus will be placed on the respective advantages and disadvantages of an inter-organisational coupling of two spatially separated training simulators (multi-located) compared to the training of SAR forces in a closed, single-located training environment.

2. OPERATIONAL BACKGROUND

The practical relevance of our research is derived from the operational practice of the DGzRS. The official duty to provide maritime SAR in the German search and rescue region of responsibility was handed over by the Federal Ministry of Transport to the non-profit organisation DGzRS. In addition to coordinating rescue operations, this organization also provides the necessary ships [2]. In Germany, the provision of air support at sea is governed by a mutual support agreement with the Ministry of Defence, which is responsible for the aeronautical search and rescue service. Consequently, SAR helicopters from the German Navy (Naval Air Wing 5) are also available on request from the DGzRS for the rescue of people in distress at sea. Moreover, in accordance with international conventions, any individuals or entities in the vicinity of a maritime emergency are obliged to render assistance. Private offshore aviation represents, therefore, a potential additional resource for the maritime rescue operations of the DGzRS. The ongoing expansion of offshore wind farms in the German Bight and the western Baltic Sea has resulted in the presence of these companies within the German Exclusive Economic Zone. The helicopters of these companies are employed to transport personnel and materials to the wind farms or back to the shore. They typically carry medical personnel. The growing economic utilisation of maritime space for the production of renewable energies consequently gives rise to an increase in the number of ships and helicopters that could potentially be deployed in rescue operations. In terms of formal regulation, the tactical interaction between these resources in a rescue situation is governed by an international set of rules. The International Aeronautical and Maritime Search and Rescue Manual (IAMSAR), published by the International Maritime Organization (IMO) and the International Civil Aviation Organization (ICAO), provides a comprehensive framework for the search and rescue of human life, and serves as the common action framework for SAR organizations in an international context [3].

In Germany, the DGzRS and the Naval Air Wing 5 utilise the IAMSAR Manual as a legally binding framework for tactical interactions. Furthermore, the officers of every commercial vessel are obliged to possess a current edition of the IAMSAR Manual vol. III in the event that their vessel becomes distressed at sea or is required to conduct search and rescue operations onsite. The situation is somewhat distinct in the context of aviation. In civil aviation, the IAMSAR Manual is not required to be carried on board, and it is largely unknown to pilots. This phenomenon has been observed on numerous occasions in the context of the operational integration of offshore helicopters in SAR missions and real-life exercises. Despite the fact that offshore helicopters have been shown to demonstrate consistently high levels of competence in the rescue of individuals in distress from vessels or from the water, deficiencies have been identified in the execution of search operations due to the absence of mandatory training requirements for helicopter pilots in this area.

In particular, dealing with major incidents at sea, such as ferry accidents (e.g. the fire on the LISCO GLORIA in the night of 8 to 9 October 2010 six nautical miles northwest of the island of Fehmarn with 235 people on board [4] when a large number of passengers and crew members have to be rescued poses particular challenges for the rescue services and requires the ad hoc coordination of rescue teams on land, at sea and in the air. Even if such incidents are fortunately rare, when they do occur, they require the rapid ad hoc coordination of a large number of emergency services which can be trained in practice only with a great deal of effort. A particular challenge here is the need for inter-organisational and even transnational planning and coordination of such real-life exercises, which can therefore only take place at relatively long intervals. While medium scale exercises lasting several days usually take several weeks to months to prepare (e.g. the SAREX exercises of the DGzRS which take place twice a year alternately in the North and Baltic Seas) [5].

Large scale exercise with a duration of up to one week usually require a year or more of preparation time. The Dynamic Mercy e.g. is an annual international LIVEX serial, led by the NATO's Allied Maritime Command (MARCOM), held for two weeks alternately in the Atlantic Ocean or the Baltic Sea with participating seagoing and airborne SAR units from ten (10) countries or more [6]. Another example was the BALTIC SAREX, an international week-long large-scale mass rescue exercise in the Baltic Sea region with dozens of vessels, several helicopters and some hundred personnel [7] – this series of exercises ended after 2015 due to the costs and resources needed to prepare and conduct this large scale exercise.

In this light and even if they can never replicate reality 100 %, simulator training seems to be a sensible alternative to real-life exercises in terms of economy and content. In the following section, we will therefore look at the importance of simulator training in maritime SAR training, its possibilities and limitations, and the role of simulator networks in creating realistic training conditions for major maritime emergencies.

3. SIMULATOR TRAINING IN MARITIME SAR

We will here concentrate mainly on the situation in Germany, but also include international aspects. In addition, we are focussing on civil solutions as access to information about military training opportunities is usually limited. Nevertheless, in many countries the maritime SAR service is more or less exclusively provided by military units [8]. Simulators play an important role in both aeronautical and nautical education and training [9].

Focussing here on nautical training, different types of simulators allow various training options, such as ship handling simulator, GMDSS simulator, VTS simulator etc. [10] [11].

Maritime simulator training aims mainly at individual hands-on training [12]. At the same time teamwork is essential in either steering a vessel [9] or an aircraft [13], so simulator-based training has to address both individual and team training.

In Germany, there are several maritime and aeronautical simulator centres with state-of-the-art technologies, however, SAR only accounts for a very small proportion of both nautical [14] and aeronautical training, is only offered as an additional qualification [15] and therefore only carried out in a highly specialised context. The possibilities for (civil) simulation-based training of maritime SAR operations are correspondingly limited. In Germany, seagoing maritime SAR personnel is mainly trained by the DGzRS at the "Seenotretterakademie" (Maritime Rescue Academy) in Neustadt/Holstein. In addition to the practical training on board, exercises in the simulator complement the qualification as a sea rescuer. For this purpose, the DGzRS has been operating its own simulator centre in Bremen since 1995.

Specialised simulation-based training for air rescue personnel is only offered at the ADAC HEMS Academy, which is also a leading international facility [17].

An international comparison shows that other sea rescue services also use simulation training for training purposes. However, the facilities available to civil SAR services in particular can only reproduce the reality of a SAR operation, especially a complex SAR situation, to a relatively limited extent, as we will discuss in section 3.2. Either interaction with other vehicles is not provided for, or only to a very limited extent, or the communication essential for the training of SAR missions is not represented, or only in a very simplified form. The following closer look at general strengths and weaknesses of simulation-based training compared to real-life exercises provides an overview of the potential of simulators for maritime training.

3.1 SIMULATOR-TRAINING: OPPORTUNITIES AND LIMITATIONS

In general, simulator-based trainings offer multiple advantages compared to real life exercises as well as some weaknesses. [10] [11] [16] [17]. The arguments listed below apply regardless of the specific training context, e.g. regardless of whether it is a general vehicle handling training or a special SAR training, which will always be more complex in practice, as not only the handling of the vehicle but also the communication and cooperation with other units must be trained. Table 1 provides an overview of the general strengths and weaknesses of simulation training, differentiated according to technical, organizational/resourced-based, and human factors aspects. However, a clear categorisation is not possible in all cases and some of these aspects touch on more than one category.

Table 1. Opportunities and Limitations of Simulator-Training

Opportunities	technical	Organisational/ resourced-based	Human factor-related	Limitations	technical	Organisational/ resourced-based	Human factor-related
Safety: Errors have no real consequences/no danger for personnel and equipment		X	X	Limited degree of immersion (winching, rescue manoeuvres) (However, can be increased through technology integration: VR/AR)	X		
Lower operating costs (lower consumption of resources, e.g. fuel, maintenance)		X		Acquisition costs, personnel costs		X	
Less planning and preparation work		X		Exercise artificiality, 'sterility': reality is partly idealized ('laboratory conditions')	X		X
Environmental friendly: Little to no negative environmental impact (very low emissions, no noise pollution)		X					
Decoupling of operational resources: no commitment of operational resources for training operations, training almost 24/7 possible		X					
High flexibility in terms of resource availability (ship types and flight patterns can be imported as required depending on the simulator equipment)	X	X					
Complete independence from environmental conditions (weather, time of day, etc.)	X						
Independence/free choice of training area	X	X					
No obstruction of sea/air traffic or no need for spatial restrictions of the exercise area		X					
Maximum plannability of training (incl. repeatability of certain situations, e.g. emergency procedures), immediate termination of situations/training possible	X	X	X				
Record and playback function increases analysis/evaluation options	X		X				
Optimised learning: • training courses can be tailored to the respective training objective • adjustment of the level of complexity and difficulty is possible Stress levels can be adjusted individually with simultaneous stress measurement: it is possible to learn in a completely stress-free manner or to specifically create stressful situations, e.g. for practising emergency procedures			X				

To summarise, despite some weaknesses, simulator-based training offers numerous advantages over real-life exercises. As far as we know, there is no convincing simulation-based alternative to the costly large-scale SAR exercises for inter-organisational training of major maritime emergencies involving air and surface rescue units. There are various reasons for this: As already mentioned above, SAR-related training content only makes up a comparatively small part of the overall nautical and aeronautical education and is generally only used as part of specialised additional training. At the same time simulator training emphasizes on the acquisition of practical, hands-on skills. Although cooperative situations can be represented in principle, they have so far only been mapped to a limited extent in comprehensive maritime SAR simulator-trainings, especially concerning the simulation of interorganizational cooperation in major incidents [18]. In individual training sessions (i.e. one individual trainee or a bridge team), cooperative situations with other units are usually simulated by the instructor using 'third-party vehicles'. Authentic communication training is possible with GMDSS simulators, which are used to train distress communication, but as these simulators are not equipped with a visual display, they cannot provide a thorough training experience and offer therefore only an insular solution [19]. Realistic cooperative simulation training can be realised through the setup of simulator networks though, i.e. the coupling of individual simulator stations as is already the case in other areas of maritime simulator training [20]. The extent to which this can also be realised for simulator-based SAR training will be examined in the following section.

3.2 ONE STEP BEYOND. EXPANDING SAR-TRAINING OPTIONS BY COUPLING SIMULATORS

In this section, we will analyse in which constellations different simulators can be connected (i.e. coupled) with each other. The focus here is less on the technical process and more on the surrounding setting. In the context of large-scale SAR exercises, this means that if a major incident is to be depicted as realistically as possible in a simulation, both flying and seagoing units (understood here as units from different domains) from different organisations should be able to train with each other. It is therefore necessary to analyse whether it is possible to combine several flight and ship simulators in such a way that the trainees can interact together in a simulation environment as authentically as possible. As this is generally only conceivable in a larger simulator centre, the question also arises as to whether it would be possible to link simulators from different, generally geographically dispersed organisations and facilities. To systematise possible coupling constellations, a distinction is thus made between 1) an intra-organisational coupling implemented within the organisation vs. an inter-organisational one between different organisations and 2) a 'domain-specific' coupling, i.e. relating only to maritime and aviation training separately vs. a 'cross-domain' coupling, i.e. the coupling creates a common training environment for both maritime and aviation personnel.

Please note that the solutions shown in table 2 do not claim to be exhaustive, but are only intended to serve as practical civil examples to illustrate the categories presented. For marine application, there are numerous examples of simulator networks, especially for military purposes. But since these are naval tactical simulators that were not developed for SAR scenarios, they will be neglected here. However, it is worth noting that the origins of creating simulator networks by coupling individual simulators date back to the early 1980s, when the SIMNET programme, developed for the US military, linked various vehicle simulators (tanks, helicopters, aircraft) for the first time, creating the first networked multi-user simulation platform [21].

In the following, a simulator network is understood as the coupling of individual simulators to a multi-user network, within which the necessary data is exchanged via suitable interfaces to be able to interact in a common simulation environment in real time.

Table 2: Categories and examples of networked simulators

	domain-specific	**cross-domain**
intra-organisational	Maritime Simulation Centre Warnemünde, MSCW (Wismar University of Applied Sciences, Rostock)	Search and Rescue Simulator Centre (DGzRS Headquarters, Bremen)
inter-organisational	European Maritime Simulator Network (EMSN)	AMARIS

In the simplest configuration, simulator coupling takes place within the organisation and on a domain-specific basis. This simplifies both the organisational and technical framework conditions and generally requires a manageable amount of coordination and development work. The MSCW, which belongs to the Warnemünde Department of Maritime Studies of the Hochschule Wismar, University of Applied Sciences, Germany as an example, accommodates several ship handling bridge simulators, which are able to simulate the steering and manoeuvring behaviour of ships of different types and sizes of vessels, ship engine simulators and Vessel Traffic Services simulation systems that are connected to a common network [20]. Neither flying SAR units nor SAR units from other organisations are integrated into this solution. With the Vessel Traffic Services simulator as an onshore unit, one could also speak of a cross-domain solution here – but since the focus here is on cooperation between flying and surface units, this aspect shall be neglected here.

In the field of flight simulators, according to our knowledge there are no civil solutions for coupled flight simulators. The HFTS NH90 simulator centre can be mentioned here as an example which operates the networking of four (4) NH90 simulators at three (3) locations. The NH90 helicopter is a military multi-purpose helicopter and is used by the Naval Air Wing 5 for SAR missions since 2023 [22]. Nevertheless, a joint simulation involving seagoing units is not included.

In order to effectively train cooperative situations in SAR missions, it makes sense to achieve this through the inter-organisational networking of SAR simulators. In this way, the possibilities for simultaneous training of a large number of trainees in a joint scenario across spatial and organisational boundaries can be significantly increased and the conditions of a major incident can be realistically simulated.

An example of the inter-organisational and transnational networking of simulators is the EMSN created as part of the EU MONALISA 2.0 project, which links 13 simulation sites in Europe and aims to improve interoperability among SAR services, passenger ships, Vessel Traffic Monitoring & Information Systems and Mission Control Centres [23]. The EMSN was not built with the primary purpose of expanding SAR training facilities, but as a platform for nautical training in general and as a test bed for new maritime applications. Nevertheless, with regard to its suitability for simulating major incidents, it should be noted that neither realistic VHF-communication could be implemented here [24] nor rescue control centres (RCC) are involved, which is essential for the effective training of SAR missions [19].

What the solutions described so far have in common is that they do not involve air units. The SAR simulator centre of the DGzRS, can be cited as an example of a network of simulators with the aim of allowing air and surface units to train together. consists of several interconnected simulator cabins – o-called cubicles – each of which represents one of the participating resources. This can be any ship or, in a special cubicle, the entire Maritime Rescue Coordination Centre (MRCC) including the coastal radio station Bremen Rescue Radio (BRR). Two variable cubicles also allow to be used as a ship, helicopter or rescue control centre as required. The variably usable helicopter stations therefore allow cross-domain interaction. Even though this is only a very simple version of a helicopter simulator, it offers pilots the opportunity for joint SAR training with ship crews in complex training scenarios. Helicopter cabins are continuously upgraded to provide greater immersion. It is also possible to integrate a further ship cubicle and two mobile operation stations (MOS). All cubicles are networked with each other and with the Instructor Centre. In contrast to all the simulators mentioned above, the networking of the cubicles offers the decisive advantage that all participants can communicate with each other in different ways (e.g., radios, telephone, and e-mail) and thus exchange information and orders. This is a fundamental prerequisite for being able to train SAR deployment procedures in a simulation at all. As soon as the MRCC cubicle is also integrated into a simulation, all essential command levels of larger SAR missions can be modelled. In addition to the communication options, the integrated cubicle for displaying the MRCC and BRR is unique worldwide. Although trainees from other organisations can train in the SAR simulator or be integrated via the decentralised MOS, there is no real inter-organisational simulator networking in this way. We are not aware of any other comparable civilian solutions. It should also be mentioned here that ships and helicopters can also train SAR situations in a joint simulation environment at the simulator centre operated by the Center for Tactics at the Naval Warfare Center in Frederikshavn (Denmark). However, to our knowledge, the possibilities for interaction between ships and helicopters are limited in various ways, for example, communication between the units is only possible via one channel so that complex SAR situations cannot be realistically modelled, nor the interaction within the air crew, e.g. while hoisting, cannot be a training objective [25].

Although the DGzRS simulator centre offers the possibility for several seagoing and flying units, including MRCC and BRR, to realistically train even larger operational situations and, above all, to represent authentic radio communication via several channels, the possibilities for a larger scenario are nevertheless restricted due to the limited number of cubicles. The helicopter simulation still only offers a very generic and therefore limited immersive flight simulation experience. The AMARIS research project launched in 2021 together with the DLR and the University Witten/Herdecke (UW/H) and further associated research partners from the aeronautical sector [26] aims to close this gap and develop a cross-domain, inter-organisational training environment with a suitable training concept.

By linking the SAR simulator centre of the DGzRS with the AVES flight simulator of the DLR, a significant extension of the current simulation possibilities can be achieved. At the same time, the coupling of generically different simulators is shown to be possible. While the SAR simulator of the DGzRS is a tactical training simulator, the AVES is a research simulator that serves – among other things – as a platform for the development and testing of new pilot assistance systems. This is not the only reason why the AVES offers a significantly higher degree of immersion compared to the variable helicopter cubicles of the SAR simulator. In the research project, external load positioning and stabilisation systems were to be tested for use in winching manoeuvres, which are particularly important for SAR missions.

The DLR itself is of course not a SAR organisation. However, as a research institution, it offers a wide range of opportunities for developing and testing solutions that can be used in practice.

In order to better understand the needs of the end users of such simulator training and to be able to take them into account appropriately in the implementation of the simulator coupling and in the development of a suitable training concept, we conducted an interview study before starting the work. This will be presented in more detail in section 4.

In the following section, both technical and human factor aspects of the coupling within the AMARIS project will be presented.

3.3 AMARIS: INTER-ORGANISATIONAL AND CROSS-DOMAIN SIMULATOR COUPLING

3.3 (a) Technical Challenges

The networks described in the previous section mainly relate to ship handling simulators, which can also be networked with each other across manufacturers. International standardisation with the High Level Architecture (HLA), IEEE1516, and Distributed Interactive Simulation (DIS), IEEE1287, provide the framework for data coupling. With regard to the simulation of major emergencies, one of the main weaknesses of existing solutions lies in the realistic implementation of adequate communication options.

AMARIS required a distributed simulation setup over the internet since the simulators are located in two distinct facilities (Bremen and Braunschweig). While AVES uses a software environment developed entirely by DLR, the DIS standard was used to link the two simulators. The SAR simulator operates as the host and is able to read and write the DIS protocol standard. A VPN tunnel was used to ensure the security of the data exchange. Figure 1 shows the setup of the coupled simulators.

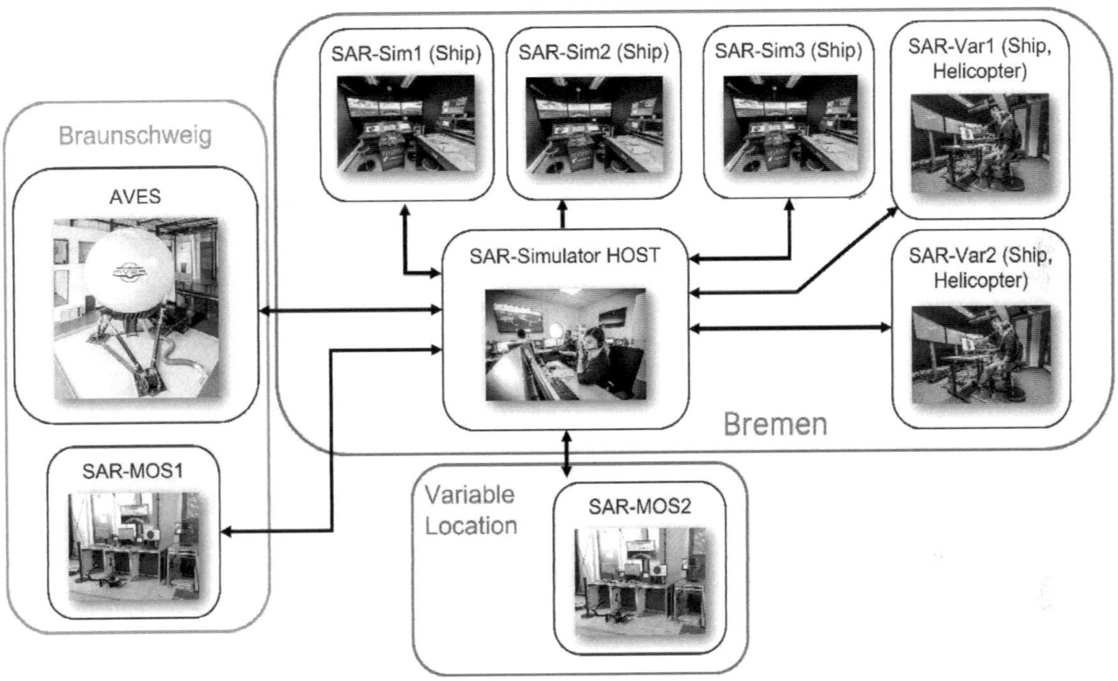

Figure 1: Simulation Setup.

Without going too much into technical details we can distinguish between two challenges that had to be addressed: First, the real-time simulation data had to be exchanged. The authentic visualisation of the environmental conditions and objects in the training scenario form an essential basis for an immersive simulation experience in the training scenario. In order to harmonise the visual representation in both simulators, the data of the objects to be simulated (vessels, aircrafts, drifting objects, winching rope, stretcher etc.) including the traffic situation and the environmental conditions had to be exchanged. For this purpose, both simulators must provide identical 3D-models of the corresponding objects so that the trainees can see the same items in both simulators. Furthermore, the environmental parameters must be specified and synchronised, i.e. the time of day and the weather conditions (clouds, rain, wind, waves) must be defined and put in an in an identical data format in order to be able to simulate, e.g. a sudden deterioration of the weather in the training scenarios. A special feature that also had to be taken into account was the consideration of the tidal range and its synchronised dashboarding in both visual systems. Asynchrony would mean that the helicopter pilots, for example, would not see the edge of the mudflats at the same coordinates as the coxswains. On the North Sea coast, however, this plays a significant role in rescue operations, e.g. when ships can no longer navigate certain fairways due to low tide. Secondly, and this was the real challenge, the communication-interface had to be implemented, i.e. a communication concept had to be created to map the real radio traffic at sea between the two simulators. Here, the AVES simulator was integrated into the already existing communication network of the DGzRS simulator network via one of the MOS. In addition to VHF communication via the international distress and calling channel 16, the AMARIS system can also realistically simulate the switch to working channels, which is essential for the coordination and communication of major incidents. This is a feature not yet implemented in comparable network solutions, as stated above.

Despite several already completed and still planned, technical enhancements (e.g. realistic fuel gauge, winching functionalities, etc the existing generic SAR simulator's helicopter models of course cannot be compared to the highly immersive simulation experience in a full-mission flight simulator but offer additional training options for helicopter pilots.

3.3 (b) Expected effects on training success

In order to ensure the success of the training in the innovative simulator environment created by the coupling, the task of our research partner at UW/H was to develop an evidence-based training concept based on current research findings on the effectiveness of simulator training. To monitor the training effect and to enable the continuous further development of the training, a research platform for behavioural team research was also to be created as part of the AMARIS project. [27] In section 3.1 we already discussed human factor related strengths and weaknesses of simulator-based training. It can be expected that with the expansion of training opportunities through the cross-domain and inter-organisational creation of a common training environment, the positive effects on learning success will also increase. As the data collection has not yet been completed, we can only present preliminary results here (see section 5).

4. METHODS

4.1 IDENTIFICATION OF END-USER REQUIREMENTS: INTERVIEW-STUDY

To create a basis for developing a coupled simulation environment based on the needs of the end users we conducted an interview study with SAR stakeholders. The interview guideline was developed by an interdisciplinary team, with input from experts in aeronautical, maritime SAR, and behavioural science. Figure 2 illustrates the design of the interview study.

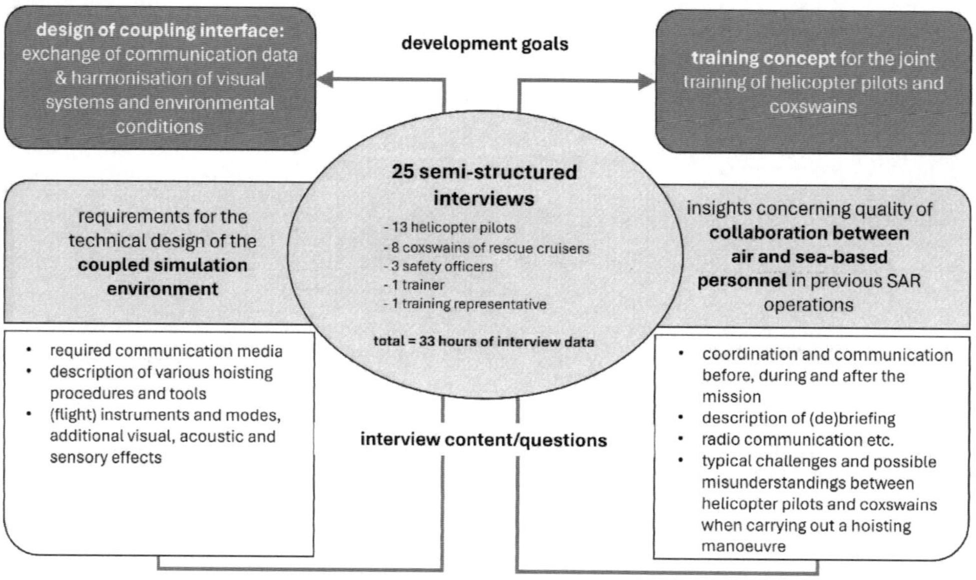

Figure 2: Design of the interview study.

Our analysis of the qualitative data drawn from the interviews revealed that although the IAMSAR Manual Vol. III [3] as international SAR standard – regulates the cooperation between surface and air units during missions – relevant regulations were not always sufficiently well known by interviewees, especially among commercial offshore pilots. Aviation and seafaring showing in part significant differences in standard operating procedures and in communication routines which can be detrimental to the quality of cooperation in both standard search procedures and hoisting manoeuvres. Although these operations are often practiced under real conditions we found out that there is only little trust in the knowledge of the working conditions of the other domain (helicopter or rescue cruiser)[1].

4.2 IDENTIFICATION OF TRAINING OBJECTIVES AND DEVELOPMENT OF TRAINING SCENARIOS

To supplement the findings from the interview study, we conducted two workshops involving various SAR experts with either aeronautical or maritime background including members of the SAR simulator trainer team and social scientists. As part of the first expert workshop, we identified training objectives and corresponding learning objectives before developing the learning objectives and the training content to be derived from gradually and with an increasing level of detail. Figure 3 provides an overview of the structure of the developed teaching and learning objectives and knowledge and skills to be taught using the example of the learning objective 'Improving communication during operations' [27].

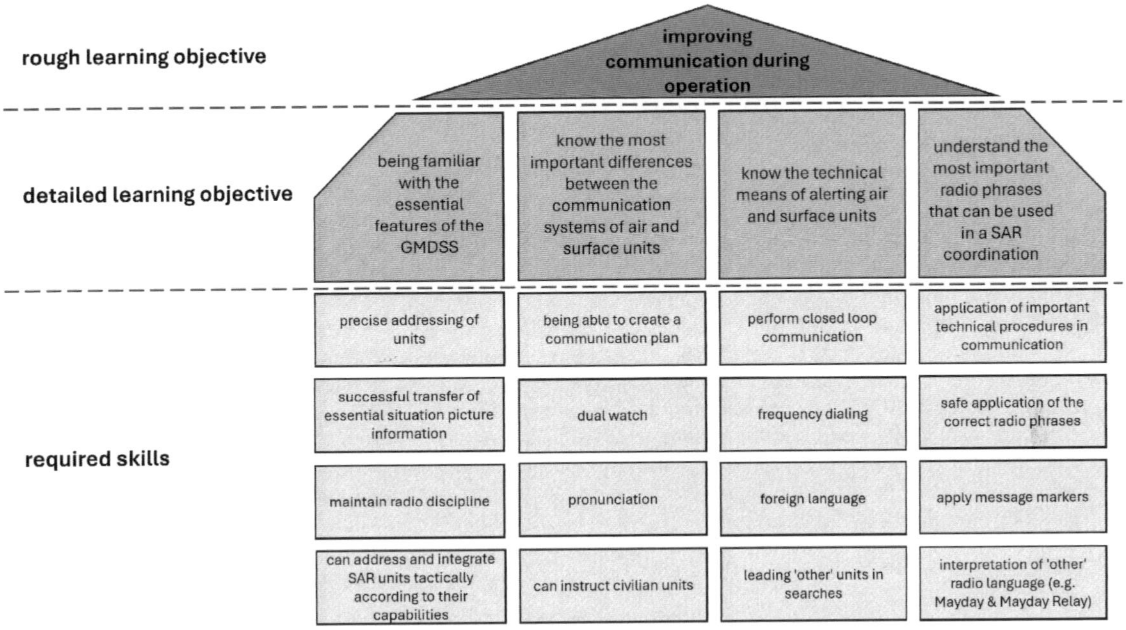

Figure 3: Learning objectives and skills.

The aim of the second workshop was to develop the AMARIS simulator scenarios. A particular focus was placed on ensuring that the previously identified learning objectives as well as the knowledge and skills were sufficiently covered by the scenario design. In interdisciplinary teams and in a multi-stage, reciprocal check-up process we developed a total of three training scenarios with increasing levels of difficulty. The scenarios encompass a range of complexities, from a basic warm-up exercise involving fundamental manoeuvres to a SAR scenario and a highly-complex major incident scenario In particular, we were able to benefit from the expertise of the DGzRS simulator training team, which has been training not only its own emergency crews but also maritime officers and captains of government vessels for over two decades, for example in On-scene Co-ordinator (OSC) trainings. As we found out in our interview-study that SAR-specific prior knowledge according to the IAMSAR manual is not equally common among all actors we provided additional written training material which was made available to the participants in preparation for the AMARIS training and covers such topics as the development of communication plans or the application of certain search patterns in SAR missions.

4.3 REALISATION OF TRAININGS

For the realisation of the courses, the scenarios were embedded in a comprehensive training concept [27]. This includes the sequence of preparatory activities such as the acquisition of suitable participants, the realisation of the training itself including the implementation of analysis tools (e.g. for stress measurement), and the final evaluation with regard to the recording of learning success and possibilities for continuous improvement of the training.

Figure 4 shows how the AMARIS training course modules rare organized, illustrating that hybrid sections via a virtual communication platform (using highly immersive conference equipment) alternate with on-site instructions in Bremen and Braunschweig. All scenarios start with a joint virtual briefing and end with a virtual debriefing session connecting all teams in Bremen and Braunschweig.

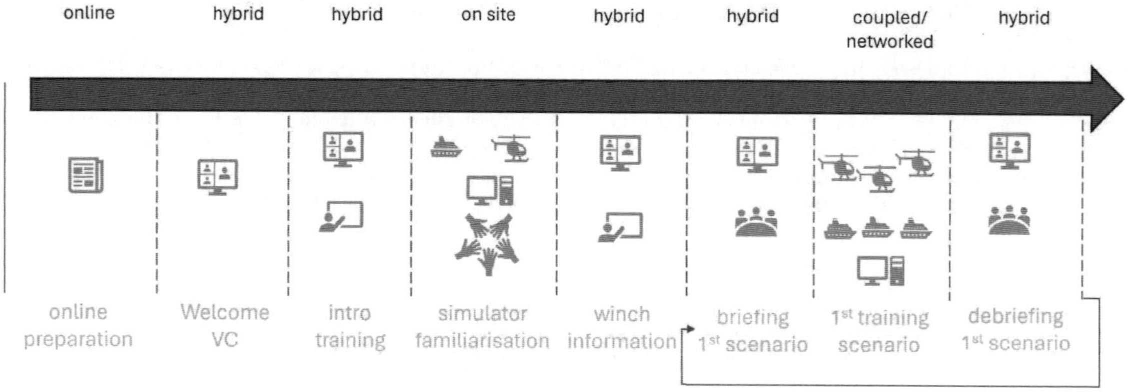

Figure 4: AMARIS training course module.

Both seafaring professionals and helicopter pilots from various organisations and authorities can take part in the training in Bremen. In the AMARIS standard setup, three cabins are used for seafaring personnel each crewed by three participants and changed in each scenario. Optionally, the two variable cabins can be staffed by teams of two helicopter pilots each.

In Braunschweig, one cubicle is available in which three HEMS professionals can train (pilot, co-pilot and hoist operator). In the default mode, therefore, nine (9) coxswains in Bremen and one helicopter crew in Braunschweig are taking part. In addition, two more helicopter crews can optionally take part in Bremen. The simulator is operated by a team of 3-4 instructors supported by 1-2 representatives from the Aeronautical Rescue Coordination Centre Glücksburg, Germany (ARCC). These individuals supervised the training from the instructor room in Bremen, simulating foreign vessels, the MRCC and the ARCC. In order to ensure the smooth running of the training and the accompanying data collection and to monitor the coupling of the two simulators, various permanent virtual meeting rooms and communication channels between the two locations were set up in the background and unnoticed by the participants. Prior to the actual test training sessions, we conducted three (3) pretest training sessions to check the technical setup, i.e. the coupled simulation environment, the timing of the training, and the various data collection tools.

4.4 DATA COLLECTION

To evaluate the effectiveness of our training concept we started to conduct a quantitative study.

The intra training data collection includes the measurement of heart rate variability (HRV) as a stress marker and thus as an indicator for measuring the level of aspiration of the simulation. HRV data is recorded using electrodes and corresponding transmitter and receiver units (*Biopac Bionomadix Portable System*) . In addition to audio data, video data of the participants is also recorded and integrated in our data collection using *NOLDUS VISO* with the *Observer XT System*. Several cameras, infrared lamps to improve the video quality and a highly sensitive microphone were installed in each cabin to record audio and video data. The video data serves as a basis for identifying interpersonal (stress) reactions with the help of suitable face reading software. Questionnaire surveys are carried out with participants and trainers before, during and after the training. The questionnaire data collection is based on the Feedback Instrument for Rescue Forces Education, which allows participants to subjectively assess their perceived training success [28]. To ensure data protection, a separate and self-contained server system with strictly regulated access rights was installed to record and store all the data collected. Figure 5 shows the data collection process during the training sessions.

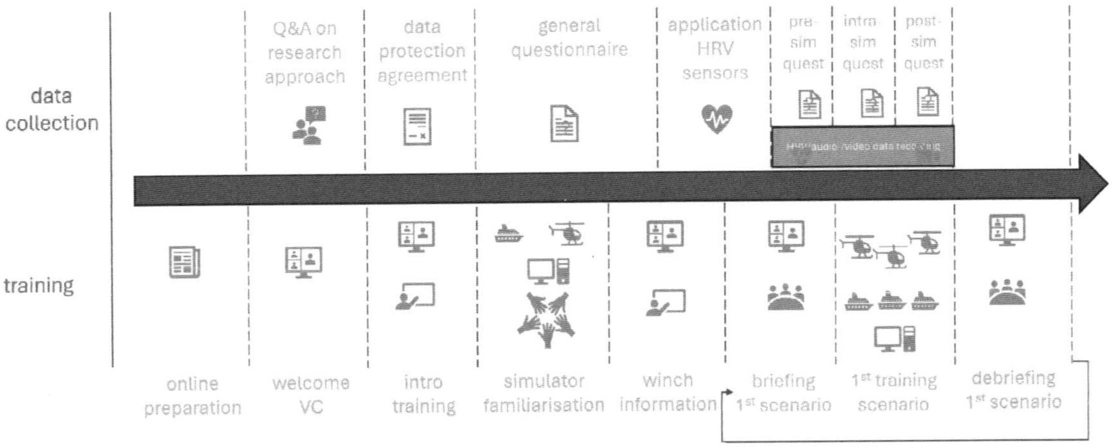

Figure 5: Data collection process.

By the end of 2023, five (5) test training sessions have been conducted to collect data. Three (3) further test training sessions are planned for the second half of 2024.

5. RESULTS

5.1 TECHNICAL DIMENSION: SIMULATOR SETUP

From a technical perspective, the coupling of the two simulators was realised through measures to harmonise the visual systems and environmental conditions as well as by implementing the communication interface. The setup was supplemented by the integration of an assistance system for helicopters, which is intended to support helicopter pilots in the winch manoeuvres essential for rescue operations at sea. This setting enables the location-independent, cross-organisational simulation of joint training scenarios for search and rescue units at sea and in the air, which is a unique selling point in international comparison. This provides trainees a completely innovative and highly immersive simulation experience that can make a significant contribution to major incident training.

Extensive programming work was required for the coupling, which cannot be described in detail here. However, it can be stated that the simulation of real-time radio traffic via the internet and the synchronisation of the position data of ships and helicopters posed the greatest challenges and required several workarounds until the final solution was implemented. Overall, however, it can be stated that the coupling was successfully completed and thus it could be shown that the networking of generically completely different simulators is also possible. Even if the number of participating units could only be increased slightly by one helicopter crew by linking with the AVES simulator, the positive experiences in the AMARIS project pave the way for the future establishment of even larger simulator networks across domains and organisations.

5.2 HUMAN FACTOR DIMENSION: TRAINING CONCEPT

Despite the large proportion of shared mental models of air and surface units, which are based not least on the common rules of IAMSAR [29], there are sometimes considerable differences in the level of knowledge with regard to the particularities of each other's search and rescue procedures and emergency communication. From the human factor point of view, we can summarise that through the AMARIS setup SAR forces are able to improve their collaboration skills and to strengthen mutual understanding of each other's working environment.

As the data collection has not yet been completed, we can only present some exemplary and preliminary results concerning the effectiveness of the AMARIS training concept. Our analysis here is limited to the questionnaire data that is already available. It will be completed once the data collection is finalised and supplemented by the results of the HRV measurements and the evaluation of the audio and video data, the latter analysed using facial expression recognition software. Using data already available, we tested the effect of the training on the acquisition and transfer of competences as rated by the participants. To identify contingencies of these training effects, we examine demographic predictors and show which group of participants felt they benefited most from the training. [1]

Figure 6 summarises key findings of the preliminary data analysis.

- 49 participants (17 permanent employees, 7 volunteers, 16 external skippers, 6 naval aviators, 2 private offshore pilots, 2 air rescuers)
- **Outcome**: to measure **perceived** training effectiveness →**self-rated competence acquisition** and **competence transfer**
- **independent variables**: **age, tenure (organization)**, **tenure (job)** →established demographic predictors of training success
- **Competence Acquisition**: **older** trainees with **limited exposure** to their **current** organization perceive themselves as particularly **benefiting** from the AMARIS training.
- **Competence Transfer**: trainees with **more extensive work experience** in the maritime or aviation environment have **more extensive mental representations** of situations to which they can relate the AMARIS training material and experiences..

Figure 6: Preliminary results: study survey

What is particularly surprising is the finding that older participants tend to benefit more from AMARIS training, as they are often less open to simulator training [30]. However, it should be noted that these are only preliminary results based on a relatively small data set.

6. DISCUSSION AND CONCLUSION

In the future, the education and training of people operating complex systems will become increasingly important. The aim of the AMARIS project was to train and improve the coordination and cooperation of sea- and air-based rescue units for complex maritime emergencies. By embedding a realistic helicopter simulation in a maritime simulation environment, aircraft and ship crews can train in a network at different locations. Providing high fidelity simulation is important [10]. For the simulation of large-scale incidents, this means integrating a large number of trainees into a scenario with the most authentic representation of radio traffic possible, as was achieved for the first time in the AMARIS project. Data collection and analysis has not yet been completed, so we can only make preliminary statements regarding the evaluation of the AMARIS training concept in terms of improved learning effects. While the objective assessment of learning outcomes in terms of individual, technical skills is relatively straightforward, the assessment of social skills, such as increasing the ability to cooperate, is a much greater challenge [31].

In the course of the first test training sessions, we also made further observations regarding the spatial arrangement of simulation training (single-located vs. multi-located). Despite the numerous advantages that result from cross-organisational simulator coupling, it must be borne in mind that the spatially separate arrangement of the simulator facilities that this usually entails also results in disadvantages that must be taken into account. Technical, organisational, resource-related and human factor-related aspects must be considered here. Due to the special circumstances of the DGzRS SAR simulator centre, where cross-domain training has been taking place at one location for around 20 years, the two solutions (stand-alone vs. coupled setup) can be directly compared with each other.

From a technical point of view, the possibility of being able to use the optimal and familiar training environment for the trainees in a coupled (multi-located) setup is offset by the significantly higher requirements for the stability of the connection and the guarantee of data security. The latter must be taken into account in particular when linking with military or governmental facilities (e.g. federal police). From an organisational/resource-oriented perspective, the single-located solution offers pooling and coordination advantages [32] in terms of training planning, personnel requirements, etc. From a human factors perspective, we observed that the personal exchange between the course participants also plays an eminent role on an informal level, which favours the superiority of the single-located solution. In contrast, the multi-located solution involves a higher familiarisation effort if the trainees (in this case pilots) have to get used to the generally unfamiliar and usually less immersive training environment. On the other hand, the possible expansion of the course programme through the linking of different facilities can be expected to have positive behavioural effects on the. participants. For helicopter pilots, there can also be a further positive effect if the training hours can be counted as flight hours if the coupled flight simulator is certified accordingly. Table 3 summarises strengths and weaknesses of the two solutions.

Table 3: Single-located and multi-located simulation networks: strengths and weaknesses.

		Technical perspective	Organizational perspective	Human factors perspective
Single-located	Strengths	• No special technical requirements • Focus on tactical simulation • Focus on communication	• No increased (cross-organisational) coordination effort for training planning • Pooling of resources possible	• Personal exchange is superior to virtual solutions even with the greatest possible immersion • Personal exchange also possible on an informal level • Social arrangements possible
	Weaknesses	• not all participants can fall back on the usual standard (helicopter training in the SAR simulator with less immersion, both for pilots and hoist operators)	• Possibly less favourable supervision ratio (trainer/participant) • Or fewer participants per course possible	• Higher familiarisation effort even with generic simulators, • additional training artificialities arise
Multi-located	Strengths	• Training with higher immersion possible • technical capabilities of the coupled simulators can be utilised equally • Domain-specific simulation expertise on site • Capability enhancement through coupling of existing resources/simulators	• Capacity expansion: more participants per course possible • No additional travel/accommodation costs for travelling participants	• Positive behavioural effects on learning success through expansion of the training programme • Pilots: Simulator flight hours can be credited if the simulator has an appropriate rating
	Weaknesses	• Greater susceptibility to technical interference factors: coupling interfaces as potential weak points • Higher requirements for data security, especially when coupling with (other) security-relevant facilities	• Special organisational requirements in training planning (coordination, availability) • Higher personnel costs: o Additional need for pedagogical personnel/specifically trained instructors o Additional need for technical staff to monitor the coupling • Duplication of non-critical resources at the respective locations	• Personal exchange is limited and can only take place during the specified slots (briefing/debriefing)

The interfaces and capabilities created by the AMARIS setup can serve as a template and standard for further cooperation and thus for the creation of training capacities for inter-agency and inter-organisational, interdisciplinary training. However, the possibilities of spatially concentrated 'single-located') solutions should also always be considered, e.g. through the cross-organisational establishment of joint simulator centres. Thus, spatially and organisationally separate 'multi-located') setups, despite numerous advantages, also pose special challenges, in particular technical and organisational ones.

From a behavioural science perspective, the new training environment created as part of the AMARIS project can be used as a platform for scientific training optimisation and team research, thus making an important contribution to the transfer of knowledge between research and practice.

7. ACKNOWLEDGEMENTS

This research was funded by the German Federal Ministry for Education and Research (BMBF) as part of the research programme *Research for Civil Security II* in the announcement *Strengthening the transfer of technology and innovation through research and development for large-scale trials, demonstration, training and further education in civil security research*, grant no. 13N15500-13N15502).

8. REFERENCES

1. LÜBCKE, T., WILHELM, H., SCHLIEBUSCH-JACOB, O., KRETER, T., MARTINI, T., and GERLACH, T., 'Inter-professional simulator training for reliable ship-helicopter interactions', *Proceedings of the 21st ISCRAM Conference,* 2024.

2. INTERNATIONAL MARITIME ORGANIZATION, 'SAR convention: International convention on maritime search and rescue, 1979; as amended by resolutions MSC.70(69) and MSC.155(78)';. *IMO publication*, 2006.

3. INTERNATIONAL MARITIME ORGANIZATION & INTERNATIONAL CIVIL AVIATION ORGANIZATION; 'International aeronautical and maritime search and rescue manual: IAMSAR Manual Vol. III: Mobile facilities', *IMO publication*, 2022.

4. BUNDESSTELLE FÜR SEEFALLUNTERSUCHUNG – FEDERAL BUREAU OF MARITIME CASUALTY INVESTIGATION, and LITHUANIAN MARITIME SAFETY ADMINISTRATION: 'Untersuchungsbericht 445/10', *AVAILABLE: https://www.bsu-bund.de/SharedDocs/pdf/DE/Unfallberichte/2012/UBericht_445_10.pdf?__blob=publicationFile&v=1*; 2012.

5. SANDER, N., 'Leben in der komplexen Lage, *AVAILABLE: https://www.seenotretter.de/magazin/sarex-such-und-rettungsuebung-der-seenotretter,* 2024.

6. NAVAL TECHNOLOGY, 'Nato's MARCOM-led SAR exercise Dynamic Mercy begins', *AVAILABLE: https://www.naval-technology.com/news/nato-sar-exercise-dynamic-mercy/?cf-view*, 2022.

7. STEIGENBERGER, N., and LÜBCKE, T., 'Space and Sensemaking in High-Reliability Task Contexts: Insights from a maritime mass rescue exercise', *Organization Studies*, 2022.

8. TAYLOR, A., and OGUTU, K., 'SAR Provision Costs and Charges in Doc 9161', *ICAO Virtual Workshop on the establishment of an effective Search and Rescue (SAR) organization, AVAILABLE: https://www.icao.int/WACAF/Documents/Meetings/2020/SAR%20Workshop/SAR%20Provision%20Costs%20and%20Charges%20in%20Doc%209161.pdf,* 2020.

9. SELLBERG, C., 'Training to become a master mariner in a simulator-based environment. The instructors' contributions to professional learning', *Gothenburg Studies in Educational Sciences 409,* 2017.

10. KIM, T., SHARMA, A., BUSTGAARD. M., GYLDENSTEN, W., NYMOEN, O., TUSHER, H., and NAZI, S., 'The continuum of simulator-based maritime training and education', *WMU Journal of Maritime Affairs*, 2021.

11. ZGHYHER, R, OSTNES, R., HALSE, K., HAREIDE, O., and JOHNSEN, E., 'Applications of Maritime Simulators in Industry and Research', *Necesse,* 2022.

12. MANUEL, M., 'Vocational and academic approaches to maritime education and training (MET): Trends, challenges and opportunities', *WMU Journal of Maritime Affairs*, 2017.

13. SCHUBERT, E., and HASLBECK, A., ‚Gestaltungskriterien für Szenarien in Flugsimulatoren zur Untersuchung von Verhalten und Leistung von Verkehrspiloten', *Der Mensch zwischen Automatisierung, Kompetenz und Verantwortung (DGLR-Bericht 2014-01),* 2014.

14. BUNDESLOTSENKAMMER, 'Die nautische Ausbildung in der Seeschiffahrt', *AVAILABLE: https://www.lotsen.de/die-wege/nautische-ausbildung-und-karriere/#ausbildungsinhalte*, 2024.

15. DRF LUFTRETTUNG, 'Der Traum vom Fliegen: Die DRF Luftrettung startet Ausbildungsprogramm für Berufshubschrauberpilot*innen. Presseinformation', *AVAILABLE: https://www.drf-luftrettung.de/hubfs/210701_Ausbildung%20Berufshubschrauberpilot_innen%20DRF%20Luftrettung.pdf*, 2021.

16- ADAC STIFTUNG, 'ADAC HEMS Academy baut ihr Angebot umfangreich aus', *AVAILABLE: https://presse.adac.de/meldungen/adac-stiftung/luftrettung/adac-hems-academy-baut-ihr-angebot-umfangreich-aus.html*, 2022.

17. ALI, A., 'Role and importance of the simulator instructor', *TransNav: International Journal on Marine Navigation and Safety of Sea Transportation*, 2006.

18. ROSSNER P, Bernhagen, M., and Bullinger, A.', Simulieren was selten geschieht. Nutzerzentrierte Entwicklung mobiler Simulatoren für komplexe maritime Großschadenslagen', *Zeitschrift für Arbeitswissenschaft*, 2021.

19. LÜBCKE, T., STEIGENBERGER, N., BORNHORST, C. and ROSSNER, P., 'Inter-organizational simulation as a training opportunity for maritime search and rescue (SAR) missions', *7th International Conference on Applied Human Factors and Ergonomics (AHFE)*, 2016.

20. BENEDICT, K, and BORNHORST, C., 'Maritime Simulation Technology for Training and Research at the Maritime Simulation Centre Warnemünde MSCW', *Proceedings of the 8th IFAC Conference on Control Applications for Marine Systems*, 2010.

21. XU, L. LIN, S., HLYNKA, A., LU, H., KAMAT, V., MENASSA, C., EL-TAWIL, S., PRAKASH, A., SPENCE, S., and MCCORMICK, J., 'Distributed Simulation Platforms and Data Passing Tools for Natural Hazards Engineering: Reviews, Limitations, and Recommendations', *International Journal of Disaster Risk Science*, 2021.

22. BUNDESMINISTERIUM DER VERTEIDIGUNG, 'NATO Helicopter 90 (NH90)', *AVAILABLE: https://www.bundeswehr.de/de/organisation/ausruestung-baainbw/ruestungsprojekte/nato-helicopter-90-nh90-#:~:text=Der%20ab%202025%20zulaufende%20NH90%20MRFH%20%28Multirole%20Frigate,im%20Schwerpunkt%20zur%20Unter-%20und%20%C3%9Cberwasser-Seekriegsf%C3%BChrung%20eingesetzt%20werden*, 2024.

23. STM – Sea Traffic Management, 'Data Exchange for safety, efficiency and environment', *AVAILABLE: https://www.seatrafficmanagement.info* , 2024.

24. JOHN, O. , BURMEISTER, H., BRÖDJE, A. , BORNHORST, C., and GRUBE; C., 'Assessing the MONALISA 2.0 Concept: Establishment of the European Maritime Simulation Network', *The 10th International Symposium 'Integrated Ship's Information Systems'*, 2014.

25. LÜBCKE, T., 'MAR-SimNET. Maritimes Simulatornetzwerk. Teilvorhaben Simulatorgestützes Training - DGzRS Schlussbericht', *AVAILABLE: https://www.tib.eu/de/suchen/id/TIBKAT:1031905987*, 2018.

26. BUNDESMINISTERIUM FÜR BILDUNG UND FORSCHUNG, 'Projektumriss AMARIS . Aeronautsche und maritime Innovationsumgebung für interorganisationale Simulationen', *AVAILABE: https://www.sifo.de/sifo/shareddocs/Downloads/files/projektumriss_amaris.pdf?__blob=publicationFile&v=3,* 2021.

27. SZCZYGIELSKI, T. GERTEN, E., and WILHELM, H., 'AMARIS Schulungskonzept', *Universität Witten/Herdec*ke, 2021.

28. THIELSCH, M., and HADZIHALILOVIC, D., 'Evaluation of fire service command unit trainings', *International Journal of Disaster Risk Science,* 2020.

29, IMSET, M., and ØVERGÅRD, K., 'Shared Mental Models of Challenging Maritime Situations: Comparisons of Ship and Shore Personnel in the Straits of Malacca and Singapore', *TransNav International Journal on Marine Navigation and Safety of Sea Transportation*, 2017.

30. GULLY, S., and CHEN, G., 'Individual differences, attribute-treatment interactions, and training outcomes', *Learning, training, and development in organizations. The organizational frontiers series*, 2010.

31. YUSHAN, P., OKSAVIK, A., and HILDRE, H., 'Making Sense of Maritime Simulators Use: A Multiple Case Study in Norway', *Technology, Knowledge and Learning*, 2020.

32. FRESE, E, GRAUMANN, M., TALAULICAR, T., and THEUVSEN, L., 'Grundlagen der Organisation. Entscheidungsorientiertes Konzept der Organisationsgestaltung', *Springer Gabler Verlag*, 2019.

9. AUTHORS' BIOGRAPHY

Oda Schliebusch-Jacob holds the current position of a Research Associate at the German Maritime Search and Rescue Service. She is responsible for carrying out various research and development activities as part of maritime security research projects. Her previous experience includes various teaching and research activities in organisational science at the faculties of economics and social sciences at the Universities of Cologne and Göttingen.

Thomas Lübcke is Head of Research & Development at the German Maritime Search and Rescue Service (DGzRS). After his studies in organizational science (MA sociology/adult education) at University of Rostock in 2006, he conducted research on high-performance teams in different domains. He joined DGzRS in 2012, where he primarily works on topics related to the management of major maritime incidents. Further work concerns the development of simulations as well as the development and integration of mission support applications, such as AI-based assistance systems for rescue operations.

AUTONOMOUS SWATH VESSELS FOR WIND FARM INSPECTIONS: A COMPARATIVE STUDY ON OPERATIONAL EFFICIENCY

S A Korkmaz, S R Turnock and **D Taunton,** Maritime Engineering, University of Southampton, UK
M Prince, Wolfson Unit MTIA, University of Southampton, UK
M Tinmouth and **N Tinmouth,** Acua Ocean, UK

SUMMARY

This study investigates the performance of autonomous SWATH vessels, both with and without foils, in offshore wind farm (OWF) operations near the Shetland Islands. Utilizing hybrid propulsion systems powered by Proton Exchange Membrane Fuel Cells (PEMFCs) and batteries, the research focuses on the vessels' operational availability during critical ROV deployment tasks. Through numerical simulations and a case study, the SWATH without foils consistently outperforms traditional monohull vessels in terms of stability and availability, particularly under challenging sea conditions. The findings highlight the advantages of SWATH vessels in maintaining consistent performance in harsh environments, crucial for the efficiency and safety of OWF maintenance. This study offers valuable insights into the operational benefits of adopting advanced SWATH vessel designs, demonstrating their potential to enhance the sustainability and effectiveness of offshore wind farm operations.

1. INTRODUCTION

As the shift to renewable energy accelerates, Offshore Wind Farms (OWFs) have become crucial for generating clean energy and meeting global demands while reducing fossil fuel dependency. Positioned to harness strong offshore winds, OWFs play a significant role in the renewable energy mix. However, their success depends on regular inspection, repair, and maintenance (IRM) operations, which are vital to prevent failures and ensure continuous operation [1]. These operations are challenging due to harsh sea conditions, including high waves, strong currents, and unpredictable weather, which complicate maintenance tasks and pose significant health and safety risks [2]. As OWFs expand into deeper waters, the need for more reliable and efficient IRM methods becomes increasingly critical.

Despite advancements in inspection technologies and optimization strategies, existing research has largely focused on operational processes rather than evaluating the operational availability of vessels specifically designed for autonomous OWF inspections [3-5]. The comparative efficiency of vessel types, such as monohulls and SWATH vessels, particularly for autonomous operations like ROV deployment, remains underexplored.

This study addresses this gap by designing and analysing an autonomous SWATH vessel powered by a hybrid propulsion system that combines Proton Exchange Membrane Fuel Cells (PEMFCs) and batteries, using liquefied hydrogen. It focuses on the vessel's operational efficiency and seakeeping performance in challenging offshore environments, offering new insights into their potential for OWF IRM operations.

A case study compares the availability of SWATH and monohull vessels at a future OWF location near the Shetland Islands, advocating for a shift from traditional diesel-powered monohulls to renewable energy-driven autonomous maritime technologies.

The paper details the research approach, including vessel design, mathematical modelling, and a case study, and concludes by emphasizing the importance of SWATH vessels in OWF maintenance, with implications for future research and industry practices.

2. REVIEW OF ADVANCEMENTS IN OFFSHORE WIND FARM MAINTENANCE AND INSPECTION TECHNOLOGIES: A FOCUS ON VESSEL DESIGN AND SEAKEEPING PERFORMANCE

The rapid expansion of OWFs has necessitated advancements in IRM technologies to ensure the reliability and longevity of these critical energy infrastructures. As OWFs are increasingly deployed in more remote and challenging environments, the need for efficient, cost-effective, and reliable maintenance strategies has become paramount. This section reviews the latest research and technological developments related to the design and seakeeping performance of vessels used in OWF IRM operations. It highlights the evolution of maintenance strategies, the integration of autonomous systems, and innovations in vessel design, particularly focusing on the role of SWATH vessels. These studies provide a comprehensive understanding of the current state of OWF maintenance, and the challenges associated with vessel performance under varying sea conditions, setting the stage for further exploration into the operational efficiency and environmental impact of these vessels.

Various studies have addressed the challenges and advancements related to OWF inspection and maintenance. For example, one study [6] emphasize the importance of risk-based maintenance planning in ensuring the reliability and extending the operational lifespan of offshore wind turbine farms. Their study highlights that effective inspection and maintenance strategies are crucial in managing the life-cycle costs and mitigating the risks associated with structural failures in offshore wind farms. Another report [7] estimate that the operations and maintenance costs could constitute up to 29.8% of the total lifecycle costs for these floating installations, underscoring the economic significance of effective maintenance strategies. A cost-benefit evaluation [8] of remote inspection technologies for offshore wind farms, demonstrating that these systems can provide significant economic benefits by reducing the need for expensive and time-consuming access to turbines. Similarly, a cost-benefit analysis [9] of robotics-driven inspections for floating OWFs, focusing on the operational expenditure and efficiency gains from using a multi-robot platform. While the study highlights the economic advantages of robotic systems in these operations, it does not delve into the specific types of vessels that would be most effective for deploying these systems.

Optimization strategies have also been a key focus in recent research, aimed at improving the operational efficiency of OWF inspections. A recent study [10] proposed a comprehensive boat-assisted drone inspection scheme for OWFs, emphasising the importance of optimal energy consumption and efficient route planning for both the drone and the supporting boat. While the study demonstrates the effectiveness of using drones for wind farm inspections, it also underscores the need for reliable boat platforms to serve as movable control centres and battery swap stations, especially in challenging offshore environments. Another research effort [11] developed a trajectory planning algorithm based on the Ant Colony Optimization metaheuristic algorithm, specifically designed to optimise the number of Autonomous Surface Vehicles (ASVs) and their routes for efficient inspection of OWFs. The study demonstrates how this approach can minimise the total travel distance and ensure complete observation of offshore structures, highlighting its potential for improving Operations and Maintenance (O&M) actions. Additionally [12] proposed an optimization approach for OWF inspection routes using a combination of K-means clustering and Genetic Algorithm (GA). Their study focuses on minimizing inspection route lengths and operational costs for inspection vessels.

One of the significant areas of research has focused on the role of autonomous and unmanned systems in OWF operations. A 2022 study [13] highlighted the potential of collaborative unmanned vehicles, including Unmanned Aerial Vehicles (UAVs), Unmanned Surface Vehicles (USVs), and Unmanned Underwater Vehicles (UUVs), for IRM operations of OWFs, underscoring the importance of reliable and efficient platforms for such operations in challenging offshore environments. Another research effort [3] further explored the use of ASVs equipped with both emersed and immersed sensing capabilities for the multi-domain inspection of OWFs. Their study highlights the efficiency of using a single ASV to generate comprehensive 3D maps that integrate data from both above and below the water surface, thereby enhancing the overall inspection process.

Another research area was about the operational challenges associated with OWF maintenance have been extensively analysed. A 2024 review [14] conducted a comprehensive review of multi-robot coverage path planning for OWF inspections, emphasizing the challenges and potential of autonomous systems in this domain Another comprehensive review [15] provides a review of monitoring, operation, and maintenance practices for smart OWFs, emphasizing the integration of advanced technologies such as digital twins and predictive maintenance to enhance efficiency and reliability. An accessibility analysis [16] conducted a comprehensive accessibility analysis for OWFs in the North Sea, focusing on the operational challenges posed by met ocean conditions for maintenance activities. While the study provides insights into the accessibility challenges for different vessel types, including workboats and helicopters, it does not specifically address the performance of vessels or the environmental impacts of propulsion systems.

Recent studies have increasingly focused on enhancing the seakeeping performance and operational efficiency of SWATH vessels in offshore conditions. One study [5] investigated the use of electric and hybrid propulsion systems in SWATH vessels, highlighting significant reductions in energy consumption and greenhouse gas emissions. The research emphasized the hydrodynamic benefits of SWATH designs, such as reduced sway and heave, which are vital for stability in varying sea conditions. Advanced simulations demonstrated improved stability and environmental impact through hybrid energy storage systems combining batteries, hydrogen fuel cells, and diesel generators. Another study [17] assessed the impact of fin stabilizers on semi-SWATH vessels, finding a notable reduction in pitch angle, enhancing stability during operations. Research on Stabilizing Ordinary Fishing Nets (SOFNets) [18] showed their effectiveness in reducing heave and pitch, offering a cost-effective stabilization method for SWATH vessels. Further investigation [19] into anti-rolling devices revealed their potential to enhance stability and reduce rolling motion in harsh sea conditions, thereby improving operational reliability. Additionally, the Operability Robustness Index (ORI) [20], though initially focused on monohulls, was introduced to optimize seakeeping performance, with principles applicable to SWATH vessels. This research underscored the importance of balancing vessel size and hull design to enhance efficiency and reduce costs. Finally, an operability analysis [21] comparing SWATH vessels with monohulls for offshore wind turbine maintenance demonstrated

SWATH vessels' superior operability in high sea states, essential for walk-to-work operations. However, it also noted higher fuel consumption due to the unique SWATH design and limited damping.

3. DESIGN SPECIFICATIONS OF SWATH AND CONVENTIONAL MONOHULL VESSEL

This section compares the design specifications of the autonomous Hybrid Unmanned Surface Vessel (H-USV) and a traditional crewed monohull vessel used in offshore operations. The H-USV, shown in Figure 1, is a SWATH vessel designed for autonomous missions, minimizing wave-induced motions crucial for safe ROV deployment. It uses a hybrid propulsion system powered by PEMFCs and batteries, with liquefied hydrogen as fuel, optimized for a speed of 3.5 knots.

Figure 1. Structural Overview of the Constructed H-USV.

The conventional monohull vessel, larger and crewed, operates at a higher speed of 10 knots, powered by two 331 kW diesel engines. While this design supports extended missions with a crew of 20 for up to four weeks, its susceptibility to wave-induced motions can impact operations, particularly during ROV deployment [22-23].

In contrast, the H-USV is engineered for long-duration, autonomous missions, with a 40 kW PEMFC and two 120 kWh lithium-ion batteries, enabling up to 40 days of operation without refuelling or crew support. The monohull excels in speed and capacity, whereas the H-USV offers superior endurance and autonomy, catering to different offshore needs. Key dimensions of both vessels are summarized in Table 1. Various remote locations on each vessel were analysed for vertical motions, velocities, and accelerations—key factors in determining the operational limits. These locations could be found in the report [23].

Table 1. Dimensions of monohull boat and SWATH [23].

Parameters	Monohull	SWATH	Units
Overall Length	38.97	13	m
WL Length	36.49	12.25	m
Beam max extents on WL	8.9	9.25	m
Draft Amidships	2	1.6	m
Wetted Area	317.6	-	m^2
Displacement	257.7	24.065	T
LCG	17.89	6.78	From zero pt. (+ve fwd) m
VCG	2	2.29	From baseline m
Kxx	3.56	3.6	m
Kyy	9.0	3.51	m
Kzz	9.0	3.51	m
Cut Water Plane Area	281.57	5.01	m^2

4. SEAKEEPING METHODOLOGY

This section outlines the seakeeping methodology employed to evaluate the hydrodynamic performance of both the H-USV and the conventional monohull vessel. The assessment was conducted using ANSYS AQWA [24], a specialized software designed to analyse vessel behaviour in response to wave interactions. Ansys AQWA employs three-dimensional radiation/diffraction theory in the frequency domain to simulate linearised hydrodynamic fluid wave loading on floating bodies, making it particularly well-suited for simulating the complex interactions between vessels and the marine environment. This software provides detailed insights into factors such as heave, pitch, and roll motions, which are critical for assessing vessel stability.

4.1 MESH GENERATION AND VESSEL MODELLING

Ansys AQWA utilizes a surface mesh composed primarily of quadrilateral panels to model the vessels. For the monohull vessel, characterized by its sharp chine, many panels were required to accurately represent its complex geometry (Figure 2 (a)). In contrast, the SWATH vessel (Figure 2 (b)) required careful consideration in the panel distribution, particularly for the hull shapes and displacement calculations. Notably, the orthogonal surfaces of the SWATH's cross deck were excluded from the model, as only the geometry below the waterline is essential for the analysis. A limitation of the AQWA meshing algorithm is that it does not account for symmetry, resulting in slight discrepancies between the port and starboard sides, which introduced minor asymmetries in the response results.

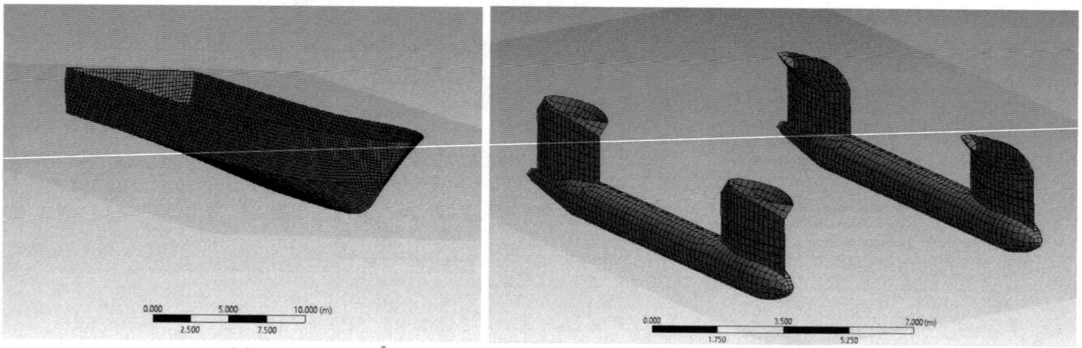

(a) **(b)**

Figure 2. Monohull (a) and SWATH (b) mesh in Ansys AQWA [23].

4.2 HYDRODYNAMIC SIMULATION AND ASSUMPTIONS

The seakeeping performance of both vessels was simulated for various forward speeds, approximated for conditions where the Froude number (Fr) is less than 0.3, consistent with limitations outlined in [24]. The monohull vessel's performance was modelled with a focus on its response to different wave conditions without additional damping corrections, reflecting the typical behaviour of crewed vessels.

For the H-USV, additional enhancements were necessary to accurately predict performance. The study incorporated additional numerical models to account for viscous damping and the lift generated by the vessel's foils, which are critical factors for the SWATH design. The method used for including these effects is based on the approach developed by [25] and more recently applied by [26] The approach used is illustrated in the flow chart in Figure 3. This iterative method updates the hydrodynamic coefficients to incorporate the effects of viscous damping and foil-generated lift, significantly reducing the amplitude of vessel motions in response to waves. These modifications were essential for providing a more realistic assessment of the SWATH vessel's stability and operational availability in various sea states.

By integrating these modelling approaches, the study achieved a detailed and accurate simulation of the vessels' hydrodynamic behaviour under various sea states. This provided essential insights into the operational reliability and stability of both vessel types, particularly in the challenging conditions of the North Sea

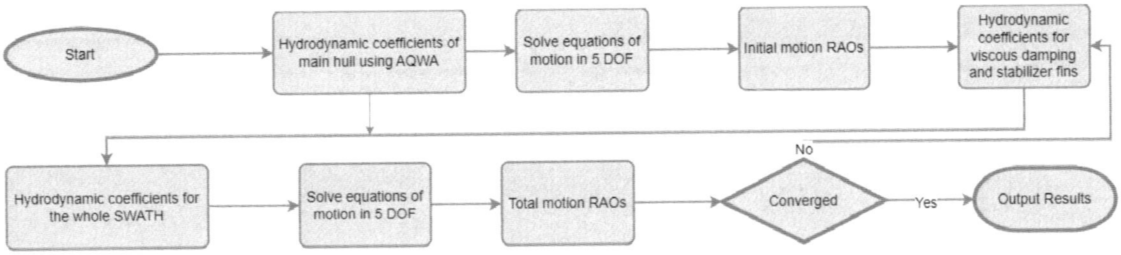

Figure 3. Flowchart showing the modelling process for SWATH [23].

4.3 SEA STATES AND VESSEL SPEEDS

The study considered a range of sea states typical of the North Sea, characterized by significant wave heights (H_s) of 1.0m, 2.0m, 3.0m, and 4.0m. These sea states were selected based on data from the NOAA Wavewatch III hindcast model, which is hourly based one year data, specifically focusing on a prospective OWF location near Shetland Islands, which represents the challenging conditions vessels frequently encounter in this region. The JONSWAP spectrum was used to model wave conditions, as it accurately reflects the steep and short-period waves that are common in the North Sea.

Tank test experiments indicated that the peak in the pitch Response Amplitude Operator (RAO) occurred at a modal period of approximately 6.5 seconds. Consequently, the final selection of wave spectra included modal periods of 4.0s, 5.0s, 6.5s, 8.0s, and 9.0s. These modal periods are critical for assessing the vessels' performance under realistic maritime conditions. Since the model is linear, the response of the vessel in any wave height can be determined by scaling the responses in 1m significant wave height by the desired wave height, though this approach does not apply to the Motion Sickness Index (MSI) and Motion-Induced Interruptions (MII) results for the monohull vessel.

Vessel speeds were set at 0 knots (stationary), 3.5 knots, and 5 knots for the SWATH vessel, and 0 knots, 4 knots, and 10 knots for the monohull vessel. These speeds were chosen to reflect typical operational scenarios during ROV deployment and recovery. The primary focus of the study was on the vessel's performance at 0 knots, particularly during stationary ROV launching and recovery operations.

4.4 LIMITING CRITERIA AND AVAILABILITY INDEX

The limiting criteria for both the SWATH and monohull vessels were derived from established literature and equipment specifications [27].

The limiting criteria for the SWATH and monohull are given in Table 2 (a) and 2 (b), respectively. For the monohull vessel, which is a crewed vessel, the launch and recovery of the ROV are manual operations. Therefore, the limiting criteria are primarily based on human performance factors such as the MSI and MII due to sliding on deck or tipping over.

For both vessels, however, there are also limiting forces associated with the ROV launch and recovery process. DNV recommended practice [28] outlines calculations to determine the forces on an ROV, but there are no limits specified.

Table 2. Limiting criteria for SWATH (a) and monohull (b) vessels [23].

Criteria	Limi
Wet Deck Slamming	Freeboard = 1.15 m Critical Velocity = 1.05 m/s
Bow Keel Slamming	Draught = 1.6 m Critical Velocity = 1.05 m/s
Foil Emergence	Draught = 1.0 m
Propeller Emergence	Draught = 1.04 m
Fuel Cell Vertical Acceleration	9.81 m/s²
Antenna Resultant Acceleration	9.81 m/s²
Mast Top Resultant Acceleration	9.81 m/s²
ROV Bottom Slamming	Freeboard = 0.7 m Critical Velocity = 1.05 m/s
ROV winch relative velocity	Critical Velocity = 1.05 m/s

(a)

Criteria	Limit
Pitch	3 degrees
Roll	8 degrees
Water on Bow Deck	Freeboard = 2.5 m
Bow Keel Slamming	Draught = 2.0 m Critical Velocity = 1.755 m/s
Propeller emergence	Draught = 2 m
ROV winch relative velocity	Critical Velocity = 1.755 m/s
Motion Sickness Index in % of crew in 120 minutes	
Motion-Induced Interruptions - Sliding	Friction = μg = 0.7*9.81
Motion-Induced Interruptions – Fore/Aft Tipping	Stance = $(L/h)*g$ = 0.17*9.81
Motion-Induced Interruptions – Side/Side Tipping	Stance = $(L/h)*g$ = 0.25*9.81

(b)

This study particularly focused on the slamming force experienced by the ROV as it is lowered to the sea surface. The slamming force F_{Slam} is calculated using the equation:

$$F_{Slam} = 1/2\ \rho C_s A_p V_s^2$$

where:
C_s is the slamming coefficient,
A_p is the horizontal projected area of the ROV $[m^2]$
V_s is the slamming impact velocity $[m/s]$, given by:

$$V_s = V_c + \sqrt{V_{ct}^2 + V_w^2}$$

where:
V_c is the hook lowering velocity,
V_{ct} is the characteristic single amplitude vertical velocity of the crane tip, and
V_w is the characteristic vertical water particle velocity, all measured in meters per second $[m/s]$.

The slamming force for a given ROV is dependent on the relative velocity of the crane winch relative to the sea surface. Without a limiting value for the force or velocity, the critical slamming velocity used for the hull and wet deck slamming is used.

The limiting criteria are used to calculate the probability of exceeding a critical value. The probability of exceedance is calculated using the Rayleigh probability distribution:

$$P_{exceed}(z > z_{crit}) = exp\left(-\frac{z_{crit}^2}{2m_{0z}}\right)$$

where:
z_{crit} is the limiting criteria,
m_{0z} is the variance of the response spectrum.

The availability index for each vessel was then calculated by determining the maximum probability of exceeding any of the limiting criteria under different sea states and speeds. This index is expressed as:

$$Availability\ Index = 100\% - \max\left(P_{exceed}\right)$$

This index represents the percentage of time that a vessel can safely operate without exceeding the limiting criteria. A higher availability index signifies greater operational reliability, which is particularly critical for the continuous, unmanned operations envisioned for autonomous SWATH vessels in offshore environments. Although this index provides an easy method of comparing the performance of the two vessels in the same conditions, it does not identify which specific criteria are the limiting factor.

By combining these criteria, the study was able to provide a clear comparison of the operational reliability of the SWATH and monohull vessels, offering valuable insights into their suitability for ROV deployment in the harsh conditions of the North Sea.

5. CASE STUDY: FUTURE OFFSHORE WIND FARM NEAR SHETLAND ISLAND

This case study aims to compare the operational performance of an autonomous SWATH vessel and a monohull conventional vessel at a prospective offshore wind farm (OWF) location near the Shetland Islands coordinates at 60.324°N 0.511°W. The location was chosen for its relevance to real-world conditions, characterized by frequent high sea states and strong winds typical of the North Sea. The study compared the operational availability of the SWATH vessel and a conventional monohull vessel, focusing on the availability of ROV launches and recoveries at the specified OWF location over one year. For this model, hourly data generated by the NOAA Wavewatch III hindcast model were utilised. Given that the study compares the ROV launching and recovery capabilities of the SWATH vessel with those of a conventional monohull vessel, the significant wave heights, wave periods, and wave directions in the selected region were analysed. The results of this analysis are presented in Figure 4 (a). Similarly, an analysis of wave directions in the selected region was conducted and is presented in Figure 4 (b). This analysis, displayed in polar coordinates, provides a comprehensive overview of the prevailing wave directions, which is crucial for understanding the operational environment and its impact on the vessel's seakeeping performance.

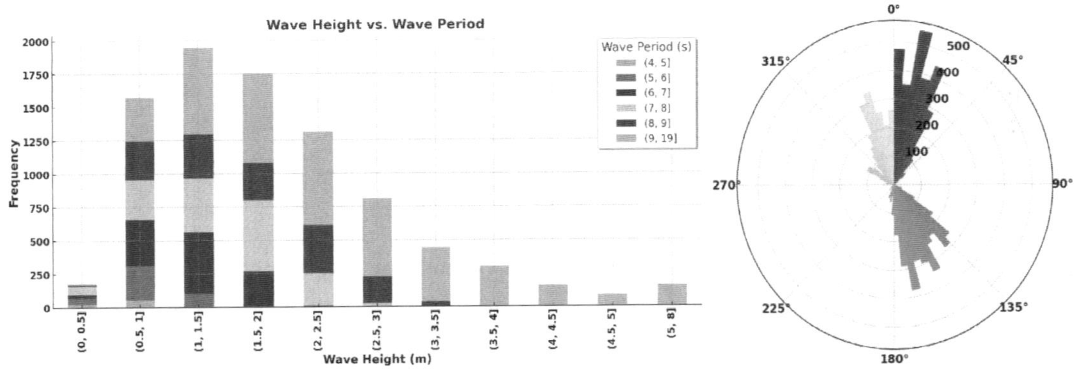

Figure 4. (a) Frequency of wave period and height combinations (b) Yearly wave directions with 5-degree bins.

6. RESULTS

6.1 OVERVIEW OF VESSEL AVAILABILITY PERFORMANCE

A comprehensive statistical analysis was conducted utilising ANSYS AQWA availability results for the vessels—Monohull, SWATH without foils, and SWATH with foils. The results offer insights into the overall availability of each vessel under varying conditions, as summarized in Table 3.

Table 3. Summary Statistics of Vessel Availability Under Varied Conditions.

Statistic	Monohull	SWATH without Foils	SWATH with Foils
Mean	74.25%	**86.58%**	84.83%
Standard Deviation	29.69%	**17.58%**	19.42%
Minimum	0.00%	**19.22%**	16.34%
25th Percentile	62.40%	**78.94%**	77.72%
Median (50th Percentile)	86.02%	**94.66%**	93.36%
75th Percentile	96.88%	**99.76%**	99.69%
Maximum	100.00%	**100.00%**	100.00%

The analysis shows that the SWATH without foils demonstrates the highest mean availability (86.58%) and the lowest variability (17.58%), suggesting that it performs more consistently across various conditions. The SWATH with foils, with a slightly lower mean availability of 84.83% and a higher standard deviation of 19.42%, indicates that the inclusion of foils introduces some variability, though the vessel still maintains high availability. In contrast, the Monohull, with the lowest mean availability (74.25%) and the highest standard deviation (29.69%), exhibits significant challenges in sustaining high availability, particularly under demanding conditions.

The SWATH vessels maintain higher minimum availability values compared to the Monohull, underscoring their better performance under challenging conditions. The median availability follows a similar pattern, with the SWATH without foils leading, followed closely by the SWATH with foils, and the Monohull lagging. The findings highlight the SWATH without foils as the most reliable vessel, particularly in operations requiring consistent performance.

In the analysis provided in Figure 5, which presents a symmetrical polar representation of vessel availability, the mean availability values are calculated across all speeds, wave heights, and wave periods for each heading. This comprehensive averaging offers a holistic view of how each vessel type performs directionally under varied conditions. We observe that the SWATH vessels, particularly the SWATH without Foils, exhibit more consistent and stable performance across all headings compared to the Monohull. This figure visually underscores the SWATH vessels' superior ability to maintain higher availability, regardless of the heading, with the SWATH without Foils excelling. This suggests that SWATH vessels are better suited for maintaining operational effectiveness across a broad range of conditions, offering a significant advantage in complex and variable maritime settings.

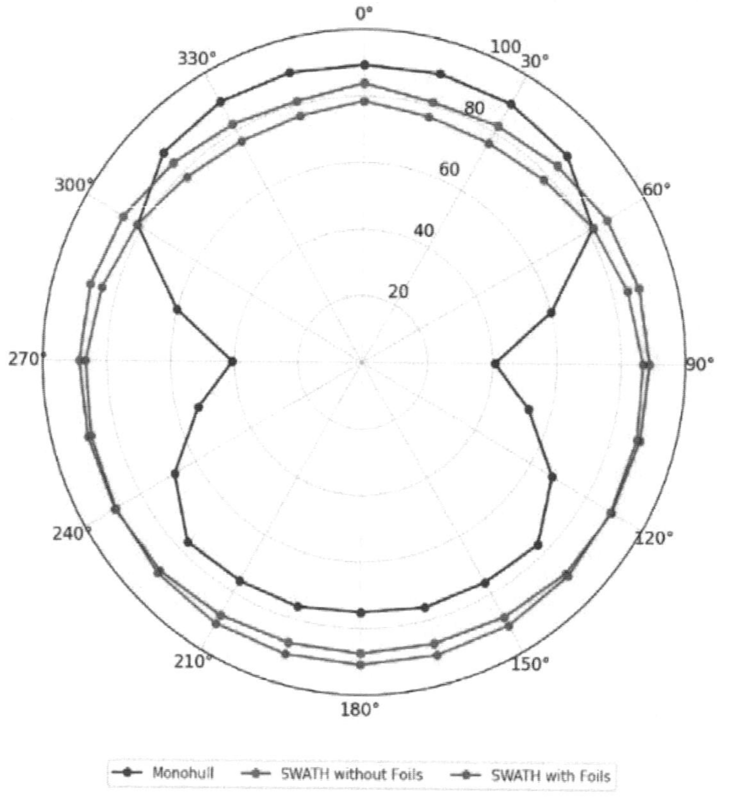

Figure 5. Symmetrical Polar Representation of Vessel Availability Across Varying Heading Angles.

Figure 6 provides a heatmap visualization of vessel availability, segmented by wave height and wave period across different speeds for each vessel type: SWATH with Foils, SWATH without Foils, and the Monohull. This figure illustrates the complex interaction between sea conditions and vessel speed on overall availability. The SWATH vessels, particularly without foils, maintain higher availability under varied sea states, as indicated by the cooler colours (blues) dominating the heatmaps. The Monohull, however, shows more red areas, indicating lower availability, especially under higher wave heights and longer wave periods. This visual comparison highlights the operational limitations of the Monohull in harsher sea conditions, further supporting the statistical analysis in Tables 4 and 5.

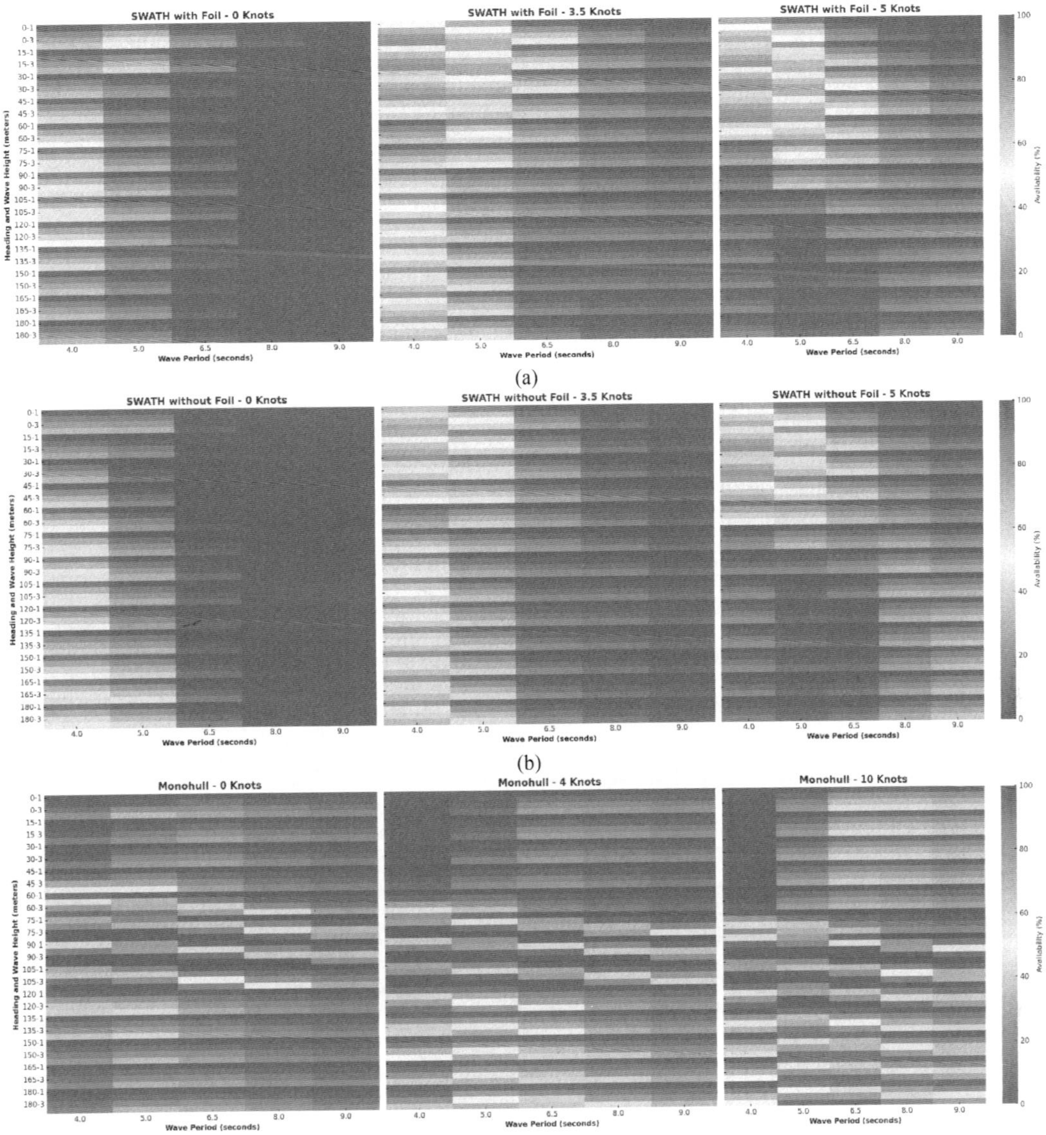

Figure 6. Availability Heatmap of Vessels (a) SWATH with Foil, (b) SWATH without Foil, (c) Monohull.

6.2 IMPACT OF SEA CONDITIONS ON VESSEL AVAILABILITY

The analysis of vessel performance across varying sea states—considering both wave height and wave period—reveals significant differences among the vessels. These conditions, collectively referred to as the "sea state," are critical in understanding how each vessel responds to the dynamic maritime environment. Table 4 and Table 5 summarize the availability statistics of the vessels under different wave heights and wave periods, respectively.

The combined analysis reveals that the SWATH vessels generally outperform the Monohull across varying sea states. Under lower sea states, characterized by a wave height of 1 meter or a shorter wave period of 4 seconds, the Monohull and SWATH vessels perform comparably in terms of mean availability. However, the SWATH without Foils leads slightly, especially in maintaining low variability.

Table 4. Impact of Wave Height on Vessel Availability Statistics.

Vessel Type	Wave Height (m)	Mean Availability (%)	Median Availability (%)	Standard Deviation (%)
SWATH with Foils	1	95.98	99.82	9.20
	2	88.44	95.76	16.44
	3	80.85	87.85	20.09
	4	74.05	79.42	22.02
SWATH without Foils	1	97.11	99.90	6.59
	2	90.09	96.85	13.83
	3	82.83	90.03	18.19
	4	76.30	82.26	20.78
Monohull	1	96.26	98.82	7.33
	2	83.18	91.43	20.35
	3	66.16	77.17	29.17
	4	51.39	60.06	32.62

Table 5. Availability Statistics by Wave Period.

Wave Period (sec)	SWATH with Foils Mean Avail. (%)	SWATH with Foils Median Avail. (%)	SWATH with Foils Std Dev (%)	SWATH without Foils Mean Avail. (%)	SWATH without Foils Median Avail. (%)	SWATH without Foils Std Dev (%)	Monohull Mean Avail. (%)	Monohull Median Avail. (%)	Monohull Std Dev (%)
4.0	68.27	70.44	24.58	69.55	71.60	22.60	70.86	89.28	34.78
5.0	76.72	81.20	21.44	81.09	85.12	17.54	65.92	79.82	33.80
6.5	89.56	95.05	12.73	92.77	97.24	9.65	70.67	82.13	29.87
8.0	94.04	98.87	7.95	94.29	99.08	8.40	79.12	87.13	24.35
9.0	95.55	99.64	6.39	95.23	99.70	8.69	84.66	91.97	19.18

As wave height increases or wave periods lengthen, the Monohull's performance deteriorates rapidly, particularly under the harsh conditions of 3 to 4 meters wave heights or wave periods extending beyond 6.5 seconds. The Monohull's availability drops to as low as 51.39% at a 4-meter wave height, with a significant rise in variability (32.62%), indicating increased unpredictability. Similarly, at a 9-second wave period, the Monohull's availability stabilizes at 84.66%, but it continues to show higher variability compared to the SWATH vessels.

The SWATH without Foils consistently shows the highest mean availability and the lowest variability across all wave heights and periods, making it the most stable and reliable vessel under varying sea states. The SWATH with Foils, while still outperforming the Monohull, demonstrates slightly lower availability and higher variability, likely due to the dynamic interactions between the foils and waves.

6.3 INFLUENCE OF VESSEL SPEED AND HEADING ON AVAILABILITY

This section examines the influence of vessel speed and heading on the availability of the two vessels—Monohull, SWATH with Foils, and SWATH without Foils. Both factors, vessel speed and heading, are critical determinants of operational effectiveness, as they directly impact the stability and reliability of the vessels under various maritime conditions.

Table 6 provide insights into how vessel speed affects availability across three aligned speed ranges: 0 - 0 knots, 4.0 - 3.5 knots, and 10 - 5 knots. At 0 - 0 knots, the analysis reveals that the SWATH vessels maintain higher mean availability compared to the Monohull. The SWATH with Foils achieves a mean availability of 91.65%, while the SWATH without Foils is slightly lower at 91.30%. In contrast, the Monohull lags with a mean availability of 78.25%. This significant difference underscores the superior design of SWATH vessels in maintaining stability and availability, particularly when stationary. Furthermore, the standard deviation for the SWATH vessels is significantly lower (13.32% for SWATH with Foils and 14.38% for SWATH without Foils) compared to the Monohull (27.80%), indicating more consistent performance for the SWATH designs.

Table 6. Vessel Availability Metrics by Speed.

Vessel Type	Vessel Speed (knot)	Mean Availability (%)	Median Availability (%)	Standard Deviation (%)
SWATH with Foils	0	91.65	99.21	13.32
	3.5	79.23	86.55	21.63
	5	83.60	91.78	20.23
SWATH without Foils	0	91.30	99.64	14.38
	3.5	83.88	92.38	19.12
	5	84.58	91.28	17.99
Monohull	0	78.25	89.01	27.80
	4	76.38	87.83	28.73
	10	68.12	79.36	31.55

As vessel speed increases to the 4.0 - 3.5 knots range, the Monohull's availability drops to 76.38%, while the SWATH vessels continue to outperform it, with 83.88% for SWATH without Foils and 79.23% for SWATH with Foils. The slight dip for the SWATH with Foils may be due to the complexity introduced by the foils at moderate speeds. The Monohull also shows the highest performance variability at 28.73%, compared to 19.12% and 21.63% for SWATH without and with Foils, respectively.

At higher speeds (10 - 5 knots), the disparity becomes more pronounced, with the Monohull's availability dropping significantly to 68.12%, much lower than the SWATH vessels (84.58% without Foils, 83.60% with Foils). The Monohull's variability remains the highest at 31.55%, while the SWATH vessels maintain more consistent performance.

Regarding heading, the SWATH vessels outperform the Monohull across all heading clusters. In the 0-15 degrees cluster, SWATH without Foils achieves the highest mean availability at 83.99%, with SWATH with Foils slightly lower at 78.55%. Although the Monohull shows a higher mean availability of 89.38% in this cluster, its stability is less consistent, with a lower standard deviation of 12.19%.

Table 7. Statistics of Vessel Availability across Different Heading Clusters.

Heading Cluster	SWATH with Foils Mean Avail. (%)	SWATH with Foils Median Avail. (%)	SWATH with Foils Std Dev (%)	SWATH without Foils Mean Avail. (%)	SWATH without Foils Median Avail. (%)	SWATH without Foils Std Dev (%)	Monohull Mean Avail. (%)	Monohull Median Avail. (%)	Monohull Std Dev (%)
0-15	78.55	88.79	24.51	83.99	95.51	21.50	89.38	94.66	12.19
15-30	76.60	86.27	25.67	81.07	92.16	23.88	89.95	95.33	11.72
30-60	77.32	86.12	24.52	83.22	92.71	20.51	89.27	93.98	12.03
60-90	83.29	88.55	17.60	87.61	93.40	14.06	70.91	84.97	32.60
90-120	87.30	92.28	14.47	88.79	93.86	14.54	46.85	47.97	38.12
120-150	89.57	95.66	14.38	89.23	96.71	14.93	72.28	81.22	26.94
150-180	90.87	97.28	14.48	87.69	95.68	16.49	76.07	86.06	25.24

As the heading shifts to 60-90 degrees, the Monohull's performance declines sharply, with a mean availability of 70.91% and a higher standard deviation of 32.60%, indicating increased variability and reduced predictability. In contrast, SWATH vessels maintain higher and more consistent availability, with the SWATH without Foils performing best at 87.61%. This trend is consistent across all headings, with the Monohull showing greater variability, particularly at larger angles. For example, at 150-180 degrees, the Monohull stabilizes somewhat with a mean availability of 76.07% but still lags behind SWATH vessels, which maintain superior performance at 90.87% (with Foils) and 87.69% (without Foils).

Overall, the combined analysis of speed and heading shows that SWATH vessels consistently outperform the Monohull, especially at higher speeds and challenging headings. Their lower variability suggests more predictable and reliable performance, making SWATH vessels better suited for operations requiring consistent availability. The Monohull, however, is less reliable under such conditions, impacting operational efficiency, safety, and cost.

6.4 RESULTS OF CASE STUDY

In this section, the comparative performance of the Monohull and SWATH without Foils (WoF) vessels is analyzed based on the availability scores derived from hourly data over a one-year period in the selected OWF region. The analysis focuses on how the availability of each vessel type aligns with the specific operational conditions experienced hourly throughout the year. It is important to note that these analyses were conducted with the vessels operating at 0 knots, meaning they were stationary during the data collection period.

The data reveals a clear distinction in performance between the Monohull and the SWATH WoF, with the latter demonstrating a significantly superior availability across all observed hours. As shown in Table 8, the average availability for the SWATH WoF is an impressive 99.85%, with minimal variability (standard deviation of 0.69%), indicating highly consistent performance. The SWATH WoF consistently maintains close to 100% availability, showcasing its robust design and capability to handle a wide range of maritime conditions with minimal disruption.

In contrast, the Monohull vessel exhibits an average availability of 96.23%, with a higher standard deviation of 4.24%. This increased variability underscores the Monohull's susceptibility to more significant fluctuations in performance, likely due to its less stable design when compared to the SWATH WoF. The Monohull's availability ranges from a low of 56.91% to a high of 99.79%, indicating that while it can perform adequately under certain conditions, it is prone to considerable performance drops under more challenging sea states in North Sea.

Table 8. Availability Analysis

Statistic	Statistic	Monohull Availability	SWATH WoF Availability
count	8737.0	8737.00	8737.00
mean	4369.0	96.23	99.85
std	2522.30	4.24	0.69
min	1.0	56.91	90.39
25%	2185.0	93.50	99.99
50%	4369.0	97.65	100.00
75%	6553.0	99.41	100.00
max	8737.0	99.79	100.00

Furthermore, the threshold analysis presented in Table 9 highlights the operational reliability of both vessels under varying conditions. For instance, the SWATH WoF consistently operates above a 95% availability threshold 99.74% of the time, whereas the Monohull only exceeds this threshold 71.36% of the time. This stark contrast in performance is even more evident when considering lower thresholds; the SWATH WoF remains operational above a 90% availability threshold 100% of the time, while the Monohull drops to 90.27%.

Table 9. Threshold Analysis

Threshold (%)	Monohull Above Threshold (%)	SWATH WoF Above Threshold (%)
70	99.86	100.00
75	99.74	100.00
80	99.65	100.00
85	99.24	100.00
90	90.27	100.00
95	71.36	99.74

These findings emphasize the SWATH WoF's clear superiority in terms of operational availability, making it a more reliable choice for maritime operations where high availability is critical. The Monohull, despite being a viable option, demonstrates limitations that could impact operational efficiency and safety, especially in less favourable sea conditions.

This data-driven assessment underscores the importance of vessel selection in maritime operations and the advantages of choosing a vessel with higher and more consistent availability, such as the SWATH WoF.

This analysis could serve as a crucial input for decision-makers when planning operations in regions where sea conditions can be unpredictable, and high vessel availability is essential for mission success. The ability to maintain consistently high availability, as evidenced by the SWATH WoF, is a critical factor in ensuring operational effectiveness and reliability in challenging maritime environments.

7. CONCLUSION

In conclusion, this study has comprehensively analysed the operational availability of Monohull and SWATH vessels, both with and without Foils, focusing on their performance under various sea states and operational conditions within a prospective offshore wind farm (OWF) location near the Shetland Islands. The research included extensive hydrodynamic analyses at 0 knots, 3.5 knots, and 5 knots for the SWATH vessels and at 0 knots, 4 knots, and 10 knots for the Monohull vessel to determine their availability across a range of operational scenarios. These analyses were conducted with respect to defined wave heights, wave periods, and wave directions to simulate realistic maritime conditions.

Although these comprehensive analyses were performed, the case study portion of this research specifically focused on comparing the vessels' performance at 0 knots, which is critical for ROV launching and recovery operations. The findings reveal a distinct advantage of the SWATH without Foils (WoF) vessel over the traditional Monohull in terms of operational availability and consistency. With an impressive average availability of 99.85%, coupled with minimal variability, the SWATH WoF demonstrates its superior design and capability to maintain high operational efficiency across varying maritime conditions. In contrast, the Monohull vessel, with an average availability of 96.23% and higher variability, shows significant performance fluctuations, particularly under harsher sea states.

In addition to the case study findings, the general availability analyses across different wave heights, wave periods, and wave directions further emphasize the SWATH vessels' superior performance. The SWATH WoF consistently outperforms the Monohull vessel across all sea conditions tested, highlighting its robustness and reliability in challenging maritime environments.

These results suggest that the SWATH WoF is better suited for operations in environments where consistent high availability is essential. The implications of this study are significant for decision-makers in maritime operations, as they underscore the importance of selecting vessels that can reliably perform under demanding conditions, thus ensuring the success and safety of offshore wind farm maintenance and inspection activities.

However, the study is not without its limitations. The analysis focused primarily on stationary operations, and future research should explore the performance of these vessels under varying speeds and more dynamic conditions. Additionally, further studies could investigate the long-term operational costs and environmental impact of these vessel designs, particularly under different operational scenarios.

In summary, this research contributes to the growing body of knowledge on vessel performance in offshore environments and provides valuable insights that could guide future developments in maritime technology and vessel design for OWFs. The comprehensive analysis of both SWATH and Monohull vessels under realistic sea conditions offers critical information that can help optimize vessel selection and deployment strategies for OWF operations.

8. ACKNOWLEDGEMENTS

This research has been supported by the Hydrogen Innovation - Future Infrastructure & Vessel Evaluation and Demonstration (HI-FIVED) project (Project Number 10057707), funded under the UK Clean Maritime Demonstration Competition Round 3 (CMDC3), and the Green Shipping Corridors: Feasibility of the world's first hydrogen-powered North Sea crossing project (Project Number 10039998), funded under Round 2 (CMDC2) of the same competition, Innovate UK.

9. REFERENCES

1. GARCÍA MÁRQUEZ, M. O., TOBIAS, A. M., PINAR PÉREZ, J. M., and PAPAELIAS M., "Condition monitoring of wind turbines: Techniques and methods," *Renewable Energy*, vol. 46, pp. 169-178, 2012.

2. VAN BUSSEL, G., and ZAAIJER, M., "Reliability, Availability and Maintenance aspects of large-scale offshore wind farms, a concepts study," *Proceedings of MAREC*, pp. 113, 2001.

3. CAMPOS, M., and COSTA, P., "Reliability Analysis of Offshore Wind Farm Components," *Ocean Engineering*, vol. 190, pp. 103543, 2021.

4. KHALID, A., and BOUZID, Z., "Cost-Benefit Analysis of Robotic Inspection in Offshore Wind Farms," *IEEE Transactions on Sustainable Energy*, vol. 13, no. 1, pp. 207-215, 2022.

5. LEBKOWSKI, A., "Innovative Techniques in Offshore Wind Turbine Blade Inspection," *Marine Technology Journal*, vol. 54, no. 4, pp. 155-169, 2020.

6. YETER, E., and OKASHA, N.M., "A Multi-Agent System Framework for Offshore Wind Farm Inspection and Maintenance," *Renewable Energy*, vol. 156, pp. 1057-1069, 2020.

7. STEHLY, T., BEITER, P., and DUFFY, P., "2019 Cost of Wind Energy Review," *National Renewable Energy Laboratory (NREL),* Technical Report NREL/TP-5000-78471, December 2020.

8. NETLAND, Ø., SPERSTAD, I.B., and HOFMANN, M., "Cost-Benefit Evaluation of Remote Inspection of Offshore Wind Farms by Simulating the Operation and Maintenance Phase," *Energy Procedia*, vol. 53, pp. 239-247, 2014.

9. KHALID, A., and ALGHAMDI, A., "Cost-Benefit Analysis of Robotics-Driven Inspection and Maintenance for Offshore Wind Farms," *Journal of Marine Science and Engineering*, vol. 11, no. 2, 2023.

10. HUANG, Y., ZHANG, J., and LI, M., "Boat-Assisted Drone Inspection of Offshore Wind Farms: A Case Study," *Journal of Marine Science and Engineering*, vol. 11, no. 3, 2023.

11. SILVA, R.J., LEITE, P.N., and PINTO, A.M., "Multi-Agent Optimization for Offshore Wind Farm Inspection using an Improved Population-based Metaheuristic," *2020 IEEE International Conference on Autonomous Robot Systems and Competitions (ICARSC)*, April 15-16, Ponta Delgada, Portugal, 2020.

12. PENG, Y., and WANG, X., "Autonomous Inspection of Offshore Wind Farms: Challenges and Future Directions," *Ocean Engineering*, vol. 274, 2024.

13. NORDIN, N., and WINDFELD, M., "Innovative Approaches to Offshore Wind Farm Inspection Using Autonomous Systems," *Journal of Ocean Technology*, vol. 17, no. 2, pp. 50-66, 2022.

14. FOSTER, T., ZHANG, X., and LI, M., "Hydrodynamic Performance of Offshore Wind Turbine Foundations," *Renewable Energy*, vol. 165, pp. 120-135, 2024.

15. KOU, K., and LAM, J.S.L., "Optimization of Offshore Wind Farm Inspection Scheduling Using Unmanned Aerial Vehicles," *Renewable Energy*, vol. 190, pp. 175-187, 2022.

16. MARTINI, R., and ZATTONI, E., "Control Strategies for Offshore Wind Turbine Structures," *International Journal of Control Engineering*, vol. 29, no. 1, pp. 45-56, 2016.

17. LUN, B.K., and ALI, A., "Analysis of Fin Stabilizer and Wave Effect on Free-Floating Semi-SWATH," *Journal of Transport System Engineering*, vol. 8, no. 1, pp. 29-35, 2021.

18. ALIAKBARI, T., ADJAMI, M., and MOONESUN, M., "An Experimental Study of Stabilizing Ordinary Fishing Nets (SOFNets) on a Stationary SWATH Ship Seakeeping Behavior Under Irregular Waves," *Ocean Engineering*, vol. 283, 115191, 2023.

19. GUAN, Y., WANG, Y., and SUN, J., "Adaptive Fault Diagnosis of Offshore Wind Turbines Based on Vibration Signal Analysis," Journal of Ocean Technology, vol. 8, no. 3, pp. 22-32, 2023.

20. GUTSCH, A., and FOSSEN, T.I., "Dynamic Positioning Systems for Offshore Wind Turbines," *IEEE Journal of Oceanic Engineering*, vol. 45, no. 2, pp. 325-337, 2020.

21. LI, Z., AND XIE, J., "Optimization of Mooring Systems for Floating Offshore Wind Turbines," *Ocean Engineering*, vol. 237, pp. 109499, 2022.

22. DASKO, D., "Feasibility of Hydrogen Powered <24m SWATH for ROV Deployment: Emissions Model Comparing the CO_2 Emissions of SWATH and Monohull on the Route from Aberdeen to Future Windfarms Near Shetland Islands," *Numerical Seakeeping Report, CMDC2-SMMI*, University of Southampton, 2023.

23. TAUNTON, D., "Feasibility of Seakeeping, Cost and GHG Efficiency of 24M SWATH Design for ROV Deployment," *Numerical Seakeeping Report, Project No: 10040003,* University of Southampton, August 12, 2023.

24. ANSYS, "AQWA Theory Manual), ANSYS, 2022.

25. LEE, C. M., "Theoretical Prediction of Motion of Small-Waterplane-Area, Twin-Hull (SWATH) Ships in Waves," David W. *Taylor Naval Ship Research and Development Center*, Bethesda, MD, 1976.

26. SUN, X. S., YAO, C. B., and YE, Q., "Numerical Investigation on Seakeeping Performance of SWATH with Three Dimensional Translating-pulsating Source Green Function," *Engineering Analysis with Boundary Elements*, vol. 73, pp. 215-225, 2016.

27. BAITIS, A., COWELL, J., CROSSLAND, P., DANIEL, A., HOLCOMBE, F., KRIKKE, M., PATTISON, J., PIGEAU, R., RICH, K., and WERTHEIM, A., "Generating and Using Human Performance Simulation Data to Guide Designers and Operators of Navy Ships: Two Large Multinational Programs," *Proceedings of RINA International Conference on Seakeeping and Weather*, 1995.

28. DNV (Det Norske Veritas), "Modelling and Analysis of Marine Operations," Recommended Practice DNV-RP-H103, 2011.

10. AUTHORS BIOGRAPHY

Suleyman Aykut Korkmaz is a Research Fellow at the University of Southampton in the Maritime Engineering Group, Department of Civil, Maritime, and Environmental Engineering. His expertise spans marine engineering, with a focus on hybrid propulsion systems, the integration of fuel cells and Li-ion batteries, the use of alternative fuels, and the decarbonization of maritime transportation.

Stephen Turnock, FRINA is Professor of Maritime Fluid Dynamics and currently Head of School of Engineering. His expertise is in the synthesis of computational, experimental, and analytical tools to maritime research challenges including maritime robotics and autonomy, performance sport, marine renewables, ship energy efficiency and decarbonisation.

Dominic Taunton is an Associate Professor at the University of Southampton. His research focuses on enhancing the design and operation of maritime vessels, particularly high-speed craft, such as semi-displacement catamarans. He specializes in experimental hydrodynamics, integrating human factors into ship design, and improving seakeeping and fuel efficiency of autonomous surface vehicles. Additionally, he has worked on performance sports engineering, adapting maritime research methods to sports like rowing and sailing.

Martyn Prince holds the position of Principal Research Engineer at the Wolfson Unit for Marine Technology and Industrial Aerodynamics at the University of Southampton. His primary areas of expertise involve aerodynamic and hydrodynamic experimental testing and vessel performance prediction.

Neil Timouth, CEO at ACUA Ocean. Neil was a founding member of the GEBCO-NF Alumni team that won the Shell Ocean Discovery XPRIZE competition for autonomous ocean mapping. Neil was also formerly the Chief Operating Officer (COO) of SEA-KIT International, a leading USV manufacturer, where he secured £3m of investment and funding, growing the company from a turnover of £50k in 2019 to over £3m in 2020.

Mike Tinmouth, Chief Operating Officer at ACUA Ocean. Mike has spent the last decade advising high-growth tech startups. BSc Law degree, major in Maritime Law and an MSc in Risk and Security Management. Mike mentors the PUBLIC GovStart accelerator, is a lecturer and business coach on the Royal Navy's innovation program, The Percy Hobart Fellowship.

STRENGTHENING CYBER RESILIENCE AND NAVIGATIONAL SAFETY IN TURKISH STRAITS: DEVELOPMENT AND TESTING OF PREVENTION STRATEGIES IN SIMULATED ENVIRONMENTS

Y B Kurt, University of Strathclyde, UK
E Akyuz, Istanbul Technical University, Turkey
R E Kurt, University of Strathclyde, UK

SUMMARY

This paper is about the research project called "Strengthening Cyber Resilience and Navigational Safety in Turkish Straits: Development and Testing of Prevention Strategies in Simulated Environments", aims to enhance maritime cybersecurity by integrating technical analysis with human factors research, focusing on Turkish Straits. With over 80% of global trade transported by sea, the maritime industry's increasing reliance on technology brings significant vulnerabilities. The study will use state-of-the-art bridge simulators and eye-tracking technology to simulate cyber-attack scenarios and capture detailed data on human responses. Additionally, the project aims to develop an AI-driven risk assessment tool to identify critical vulnerabilities and guide the creation of effective mitigation strategies. The anticipated outcomes include improved cyber resilience for maritime operations, enhanced training programs, and policy recommendations for national maritime authorities and shipping companies. This comprehensive approach will significantly contribute to the field of maritime cybersecurity, benefiting both the Turkish Straits and global maritime navigation.

1. INTRODUCTION

The maritime industry, especially in transportation, is fundamental to global trade, as it enables the efficient and reliable transfer of a wide range of products, from raw materials to finished goods. More than 80% of all goods traded globally are transported by sea and it is estimated to grow more between 2024-2028 [1, 2]. Its operations involve a complex interconnection of shipping routes, ports, and logistics services that are essential for maintaining the flow of trade between countries. Consequently, the maritime industry has been swiftly integrating technological advancements to enhance the safety, and efficiency of its operations. In recent years, dependence on technology has surged with the arrival of advanced monitoring systems, automation and autonomous ships, new propulsion systems, and the adoption of alternative fuels. As a natural consequence of this, the industry is network-centric, with a large number of independent players [3].

However, this increasing reliance on technology also presents new challenges and vulnerabilities. Digital transformation challenges in the maritime industry covers several critical areas: data security, lack of standardization, technology selection, skills and talent shortages, and organizational inertia and resistance to change [3]. Thus, concerns are growing, particularly in the area of maritime cyber security, as the increasing reliance on digital systems and automation in maritime operations exposes vulnerabilities and potential risks. Cyberattacks against commercial ships have the potential to permanently damage ship systems, disrupt with external communication, and obtain confidential company information, all of which put personnel and navigational safety in danger and damage the companies' reputations [4]. Shipping companies and vessels are continually becoming targets for a diverse array of cyber-attacks. It has been discovered that between 2017 and 2023, cyberattacks on important ports rose by 900%, to a daily average of over 100 acts [5]. In 2023, 14% of maritime industry paid ransom to regain access to their systems, rising from 3% in 2022 [6]. Recently, ships in war zones, such as the Eastern Mediterranean and Black Sea regions, are increasingly encountering GPS spoofing and jamming incidents [7]. Experts widely recognize that the frequency of cyber-attacks in this area is set to rise [8].

This is particularly essential to the Turkish Straits, one of the world's most important shipping routes, and only route that connects the Mediterranean to the Black Sea. The Turkish Straits, comprising the Istanbul Strait, the Sea of Marmara, and the Çanakkale Strait, are among the most critical maritime routes globally. This region sees the passage of 50000 of vessels annually, and 20% of them are tankers, including those carrying hazardous cargoes such as crude oil, liquefied gas, and chemical products [9]. Given the high traffic density and the presence of dangerous goods, the potential impact of maritime accidents in the Turkish Straits is significant. Turkish Straits region, home to over 20 million people, are especially vulnerable to cyberattacks in the event of maritime accidents [10, 11]. In recent years, the digital transformation in the maritime industry has brought about new vulnerabilities, particularly regarding cybersecurity. Maritime organizations must adapt to these evolving threats as cyber-attacks on vessel systems and sensors, such as spoofing, jamming, malware, and denial of service, are becoming increasingly sophisticated. Such hacks can cause rapid navigational and operational problems, data theft, and physical damage, which can affect ship operators' safety, the economy, and the environment. Hackers may gain control of a ship's Integrated Bridge System (IBS) and operate it as intended. These vulnerabilities go beyond the IBS, including potential unauthorized access and control of the ship's propulsion/rudder and ballast water management systems. An attack that occurs in a critical waterway will have a significant impact. The Ever Given shut down the Suez Canal for six days in 2021, disrupting the world supply system [12]. Recognizing these dangers is essential for creating an effective maritime infrastructure cyber defence strategy. The current systems in place in the Turkish Straits may have limitations in fully addressing potential cyber-attack risks. Cyberattacks on ships transiting the Turkish Straits could harm the local community, fishing, tourism, and canal transportation. Malicious parties will continually attempt cyber-attacks as they search for new methods and most of the time it is difficult to identify anomalies or cyberthreats in the sophisticated, technologically advanced maritime environment [13]. Thus, there is need for resilient systems to ensure robust security. Although often regarded as primarily a technical issue due to factors like networks and software, the entire problem should be addressed holistically, considering that humans are integral parts of the overall system and its operations. Human factors can significantly influence both vulnerabilities and safeguards in maritime operations. Crew members and port employees can either inadvertently become victims of cyber threats or play a crucial role in defending against them. Given that this is a relatively new field, the absence of guidance from regulatory bodies has made problems with human factors with navigational devices a barrier to safe and efficient maritime operations [14].

Therefore, it is crucial to assess seafarers' knowledge, awareness, and responses to cyber-attacks, as well as to understand how humans respond in various situations when exposed to such attacks [15]. Because of swift digitalisation, the maritime industry faces significant challenges, including evolving cyber threats, the need for seafarers to develop new skillsets, and the necessity for the context and structure of learning to adapt to operational changes. [16]. Thus, greater emphasis should be placed on human behaviour in the context of maritime cyber security, along with an understanding of how to support those working within the maritime sector [17]. However, human-factors (HF) and human-centred design (HCD) are often overlooked in maritime education, which typically places more emphasis on technological aspects [18]. However, these concepts require further in-depth examination to enhance cyber-resilience effectively. The term cyber-resilience refers to the capacity to foresee, endure, recover from, and adjust to harmful circumstances or attacks on cyber systems, ensuring that mission goals reliant on these systems can be met even in a challenging cyber landscape [19]. It is an ongoing concept that covers governance, risk management, and corporate culture, all of which must be continually updated.

The primary objective of this project is to significantly enhance the shipping industry's preparedness for cyber threats, particularly focusing on modern ships navigating the Turkish Straits. By implementing advanced risk modelling and simulation-based training, the project aims to equip maritime operators with the necessary tools and knowledge to anticipate, mitigate, and respond effectively to potential cyber incidents. This proactive approach will help safer and more secure maritime operations, safeguarding both the vessels and the vital economic and environmental interests associated with the Turkish Straits.

This paper is structured as follows: The Section 1 provides an overview of the current landscape of maritime cyber security and the significance of addressing both technological vulnerabilities and human factors. The Section 2 examines existing research and identifies gaps in the current understanding of maritime cyber threats, including the challenges posed by technological advances and the role of human factors. The Section 3 details the research design, data collection, and analysis methods planned for the study, with a focus on AI-based risk modelling. The Section 4 interprets these expected findings, exploring their potential impact on maritime operations and future research directions. Finally, Section 5 summarizes the key points of the paper and suggests directions for future research, emphasizing the project's contribution to enhancing cybersecurity resilience in the Turkish Straits.

2. LITERATURE REVIEW

The maritime literature contains numerous studies on cyber risk assessment, human factors, and resilience strategies. Researchers and practitioners, similarly, have recognized the critical need to address these challenges comprehensively. In this review, we explore key approaches and findings across these areas, synthesizing insights from a wide range of sources to provide a holistic understanding of the current state of maritime cybersecurity.

From a regulatory wise, several recommendations and regulations have been issued on the topic. The International Association of Classification Societies (IACS) has issued Unified Requirements E26 and E27[20, 21], focusing on the cyber resilience of ships. These requirements aim to guarantee the secure integration of both Operational Technology (OT) and Information Technology (IT) systems into a vessel's network throughout its design, construction, commissioning, and operational phases. The guideline covers five critical areas: equipment identification, protection, attack detection, response, and recovery and requirements are mandatory for all new vessels contracted for construction on or after 1 July 2024. Additionally, The Maritime Safety Committee adopted Resolution MSC.428(98) [22], urging administrations to ensure that cyber risks are adequately integrated into current safety management systems.

In addressing institutional challenges and strategies, Liebetrau et al. [23] argue that maritime counter-piracy operations off Somalia and maritime cybersecurity governance offer valuable insights for the institutional challenges in critical maritime infrastructure protection (CMIP). The article evaluates three European institutional design processes: NATO's transatlantic approach to integrate CMIP, the EU's efforts to enhance CMIP in European sea basins, and Belgium's initiative for the North Sea. The evaluation shows NATO's involvement in institution building, particularly with industry relations, and the EU's broad inclusion of CMIP in various initiatives, indicating multiple forms of institution building. Further exploring institutional frameworks, Drazovich et al. [24] identified two major deficiencies in the current maritime cybersecurity guidelines: they fail to offer holistic recommendations for key stakeholders in the maritime transportation system, and they are not sufficiently based on rigorous research. Furthermore, Drazovich advocated for the inclusion of directives aimed at enhancing resilience, such as a detailed cyber intrusion response and recovery plan, and recommendations for implementing redundant systems to bolster cyber resilience. By analysing current cybersecurity dynamics, Hopcraft et al. [25] highlighted the critical need for a comprehensive cyber code to ensure effective cyber resilience.

In the realm of cyber risk assessment methods, Yousaf et al. [26]'s research focused on improving the FMECA-ATT&CK cyber risk assessment method, combining FMECA and the MITRE ATT&CK framework. They adapted FMECA-ATT&CK specifically to assess cyber risks in AI systems on autonomous ships, such as engine monitoring and control. Key findings from applying this approach to a model of an autonomous cargo ship identified the Navigation Situation Awareness System (NSAS) as the highest risk. Similarly, Uflaz et al. [4] proposes a probabilistic method for assessing cyber threats in maritime bridge navigation systems. The study uses a Bayesian network model to evaluate cyber risks for integrated bridge systems. Marine security experts assessed 32 threats using FMECA parameters, combined with Dempster-Shafer theory. The research concluded that AIS spoofing represents the most significant risk among the threats analysed. Park et al. [27] mentioned that to evaluate the tactics that reduce risks and identify the most economical ways to enhance cybersecurity and resilience in the maritime sector, more study is nevertheless required.

Bolbot et al. [28] performed a comprehensive bibliometric analysis of maritime cybersecurity literature. The study identified key research topics, methodologies used, and primary challenges and future directions in the field. Results revealed that the main focus of academic research in maritime cybersecurity has been twofold: 1) developing and applying cybersecurity risk assessment techniques, and 2) designing monitoring and intrusion detection systems for maritime cyberattacks. Their analysis provides a roadmap for future research and development in maritime cybersecurity. Erbas et al. [29] conducted a comprehensive review of published literature to systematically analyse and compare current approaches to threat modelling and risk assessment in maritime cybersecurity. The study also pinpoints significant challenges in this field.

When considering human factors in cybersecurity, Nganga et al. [30]'s study examined how specific human factors in the maritime domain affect M-SOC analysts' ability to adapt and respond to cyber threats against ships. The researchers used interviews and thematic analysis to gather and interpret data. Their main finding highlights the crucial role of vessel operators in enhancing M-SOC analysts' adaptive responses. Erstad et al. [31] Erstad emphasizes the importance of studying human factors in maritime cybersecurity, noting that humans are at the forefront during potential cyber emergencies at sea. The research defines cyber resilience in a maritime context as a nautical system's ability to maintain and adjust its operations while simultaneously anticipating, recovering from, and evolving in response to cyber threats, all in the most efficient timeframe possible. Dominguez-Péry et al. [32] examined at how cybersecurity hygiene and the culture of maritime safety affected shipboard safety, particularly in terms of the resilience and collaboration of maritime professionals. The empirical results of their quantitative study showed that maritime professionals on ships with a strong safety culture and enhanced cybersecurity awareness considerably enhanced the ships' overall resilience capabilities. Furthermore, it was found that stakeholder collaboration was positively impacted by efficient onboarding decision-making. Soner et al. [33] states that cyber resilience depends on skilled seafarers because tasks requiring a high level of cybersecurity knowledge, comprehension, and skill increase the likelihood of human error. Additionally, they mentioned there is a critical relationship between potential human error and information-sharing policies because cross-organizational information-sharing is essential for cyber resilience because cyberattacks are often stealthy and difficult to detect. There are studies that utilises questionnaire method with seafarers. Haugli-Sandvik et al. [34] employed a quantitative approach to examine deck officers' perceptions of cyber risks associated with both IT and OT systems in maritime environments. The key finding was that cybersecurity training and firsthand experience with cyber-attacks specifically influenced deck officers' risk perceptions related to operational technology. Similarly, Palbar Misas et al. [35]'s research aimed to understand SA's importance and the challenges it faces in remote operations by using a multi-method approach such as tabletop exercises and questionnaires. The study's findings indicate that as maritime operations evolve, future navigators will require new training programs.

In the context of cyber resilience, Collier et al. [36] addressed the topic from the perspective of how non-compliance of microelectronic manufacturing technology with cyber-physical security measures can impact the resilience of cyber-physical systems and, consequently, the supply chain. Onishchenko et al. [37] identified the processes for the development of cyber resilience monitoring and response plan. Additionally, revealed that modern companies, including those in the maritime industry, are largely unaware of how to protect against cyber threats and the risks of confidential information leaks, with unintentional crew actions identified as the most significant vulnerability. Potamos et al. [13] argues that integrating navigation and surveillance systems, classified as maritime operational technology (OT) sensors, can enhance cyber situational awareness in maritime environments. Their study explores whether combining OT data, typically used for general maritime awareness, could also boost real-time or near-real-time detection of cyber incidents. The research includes a comprehensive review of relevant literature, with a focus sensors and data fusion techniques that could strengthen maritime cybersecurity. Khandker et al. [38]'s research aimed at improving cyber-resilience from a technical perspective by implementing and assessing 11 various tests/attacks on 19 AIS configurations within a controlled environment. The aim was to improve the early detection of emerging vulnerabilities in AIS to enable effective mitigation of AIS attacks in the near future. The findings revealed that about 89% of the configurations were susceptible to Denial-of-Service (DoS) attacks at the AIS protocol level. In examining papers outside the maritime sector, Rajamäki et al. [39] suggests that cybersecurity crosses traditional risk management to become a shared responsibility among leadership, ICT staff, and other personnel, asserting that organizations can significantly enhance their resilience by investing in the education of their people, processes, and technology.

While the current body of literature includes numerous papers on cyber security, reflecting its status as a hot and increasingly popular topic, these studies tend to cover a broad range of aspects. However, they fall short in providing in-depth analyses of the specific vulnerabilities introduced by modern technological advancements and the interconnected nature of maritime operations in narrow waters or straits. Notably, there is a conspicuous absence of focused research on the unique challenges posed by the Turkish Straits. The critical importance of this region in global maritime traffic underscores the need for targeted studies that examine its specific cyber security vulnerabilities, thereby filling the existing research gap and contributing to a more robust understanding of maritime cyber threats in such strategic waterways.

3. METHODOLOGY

It is essential to identify the risks and gather data in order to analyse maritime cyber-security risks in congested waterways.

For this reason, the research design for this project employs a mixed-methods approach that integrates both quantitative and qualitative data to achieve a comprehensive understanding of maritime cyber threats. Figure 1 shows the workflow of the project. In the quantitative approach, the data from simulator experiments and eye trackers will be analysed, while in the qualitative approach, interviews with seafarers and post-experiment surveys will be carried out. Consequently, the research aims to not only identify critical points of vulnerability but also to explore the resilience of maritime systems against cyber threats. The research will address the following questions, aligned with the project's objectives:

RQ 1: How do seafarers respond to simulated cyber-attacks in a ship bridge environment?
RQ 2: What is the impact of cyber security training and awareness on seafarers' performance during a cyber-attack?
RQ 3: How do human cognitive factors such as workload, stress, and attention vary during a cyber-attack scenario?
RQ 4: What is the effectiveness of using wearable technologies to study human reactions during cyber-attack scenarios on board a ship?

By answering these questions, the study aims to deeply understand the human and technical sides of maritime cyber security, which will help make maritime operations safer and more resilient.

When it comes to high-risk operations, training simulators that can offer cost-effective and safe working environments are an essential part of the operators' education [40, 41]. By enabling repeated practice and immediate feedback, simulators enhance the learning experience and improve operational readiness. Additionally, simulation experiments serve as a crucial component of data collection[42-44]. For this project, the experiments will be conducted using a state-of-the-art bridge simulator, which is an advanced computer system designed to replicate ship operations, port settings, weather conditions, and traffic situations for evaluation, and research purposes. It is equipped with sophisticated monitoring and control systems that creates a realistic and immersive testing ground for various cyber-attack scenarios such as GPS spoofing, jamming, and malware intrusions. By replicating these scenarios in a controlled environment, the study aims to gather detailed data on the operational impacts and the effectiveness of different defence mechanisms.

In addition to the use of simulators, the wearable technologies will be utilised. Eye tracking technology monitors and records eye positions and movements, offering valuable insights into human attention patterns, intentions, and behavioural responses. Eye tracking technologies have been applied in various industries to assess both current and future work environments, particularly those with high levels of risk [45]. About the user's visual and attentional processes, the eye tracker provides objective, quantitative insights. It allows interactive systems responding to user behaviour based on observed eye movements, such as saccades, fixations, gaze paths, eye blinks and pupil dilation, making it an effective input device for a variety of visually mediated applications [46]. This technology has previously been utilized in experiments involving seafarers in simulator environments [45, 47, 48]. Eye-tracking technology will be integrated into the simulation environment to assess human reactions and will capture detailed data on operator behaviour, visual attention and decision-making processes, allowing the study to identify key factors that influence the effectiveness of human responses to cyber threats. Following this, the collected data will be integrated into the human-in-the-loop assessment that is focusing on evaluating the performance and response of human operators when faced with cyber threats.

The post-experiment data analysis phase will contain converting raw data from simulator experiments into interpretable results. This stage includes systematic examination of data to uncover patterns, trends, and key performance metrics that illuminate human interactions and responses under various cyber-attack scenarios. Initially, preliminary data will be cleaned to remove inconsistencies and errors to ensure uniformity and comparability across datasets. Analysis will focus on identifying recurring patterns within the data by examining operator behaviours and responses to cyber threats. It is intended to provide insights into response strategies, areas of uncertainty, and overall operator performance.

In the meantime, key performance metrics will be defined and extracted to quantitatively measure operator performance that include reaction times, response accuracy, frequency and duration of eye fixations on critical interface elements, and action sequences during cyber-attack scenarios. The key performance metrics obtained will provide a solid quantitative foundation for assessing the effectiveness of the training programs and the ability of operators to manage cyber threats.

Following identification of patterns, trends, and key metrics, initial findings will be evaluated to understand broader implications. Each simulation will be configured to mirror real-world conditions in the Turkish Straits. The data gained from this analysis will inform refinement of the AI-driven risk assessment model. Integration of detailed human performance data and understanding of operator response nuances will allow model adjustment to better predict and simulate real-world scenarios. The insights gained will be used to evaluate and refine both the AI models and the human-machine interface (HMI) elements. Observing human interactions allows for adjustments to improve the AI algorithms' predictive accuracy and enhance the HMI's usability. After the assessments, a detailed debriefing sessions will be held with participants to collect qualitative feedback, which will help us better understand how humans interact with the system.

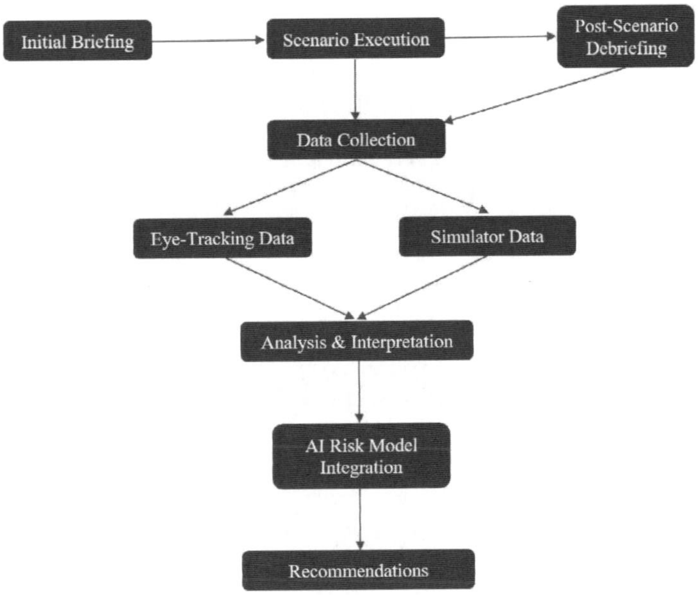

Figure. 1 Project Workflow

3.1 IMPLEMENTATION PLAN

The experimental procedure will follow a structured approach consisting of three stages: an initial briefing, scenario execution, and a post-scenario debriefing, seen in Figure 2. This methodology follows the guidelines of IMO Model Courses 6.09 to ensure that the simulator-based experiment is comprehensive and systematically evaluated. [44, 49].

An open call will be issued to seafarers, including cadets, who have been working on ships and are willing to participate in the experiments. Experiments will be conducted with participants who volunteer to take part. The briefing phase begins with participants receiving an ethics statement and consent forms to secure they are fully informed and participating voluntarily. Following this, participants complete a demographic questionnaire, providing essential background information that may influence the study's outcomes. This phase also includes an extended familiarization session with the bridge simulator. During this session, participants will get acquainted with the simulator's controls and interface to help them feel comfortable with its functionalities. They practice various standard operations to build confidence and make sure that their responses during the simulation phase reflect the experimental conditions rather than any unfamiliarity with the simulator. The scenario execution phase comprises a baseline period, attack simulations, and data collection. Data collection for this project will focus on human performance metrics captured through eye-tracking devices and will enable real-time monitoring and analysis of user interactions. The primary participant will be equipped with eye-tracking devices to monitor their eye movements. Additionally, their actions and communications during the exercises will be recorded for evaluations at a later stage. The portable eye tracker will synchronously capture the scanning and gazing patterns of the subjects and provide detailed data on the duration, frequency, location, and sequence of their visual focus on the display. During the initial three minutes, participants operate in a scenario without any targets. This period will serve to establish a control for their natural behaviour and performance. Following the baseline period, participants encounter cyber-attack scenario. Throughout this phase, detailed data on participants' actions, communications, and visual attention will be collected. Upon completion of the scenarios, open-ended discussions will be conducted with the participants during the debriefing session to gather qualitative insights about their experiences. A subjective method, NASA Task Load Index (NASA-TLX), will be used to assess perceived workload. It evaluates mental workload across six dimensions: mental demand, physical demand, temporal demand, performance, effort, and frustration [50]. Feature extraction will be conducted to identify and isolate relevant attributes essential for risk modelling and human-machine interface (HMI) assessment. Eye-tracking data will be segmented to detect patterns in user interactions, which will inform the development of the rule-based AI model.

Training the rule-based AI model will involve developing and refining rules derived from domain expertise and historical data analysis. Eye-tracking data will play a key role in this phase by providing information on common user behaviours and where users focus their attention during system interactions [51]. These data will help refining the rules to better align with actual user behaviours and enhance the model's accuracy. Additionally, the model will incorporate HMI variables such as interface complexity, usability scores, and user interaction patterns, ensuring that human factors are comprehensively considered in the risk assessment process.

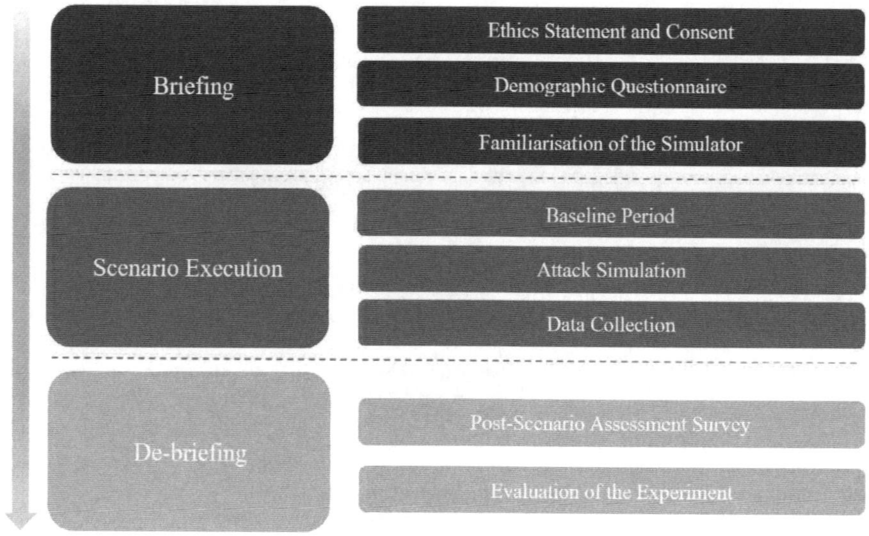

Figure. 2 Experiment Design

4. EXPECTED OUTCOMES AND DISCUSSION

The purpose of this research is to provide an in-depth analysis of cyber threats to maritime operations by combining technical analysis with human factors research. There is expected to be a wide range of outcomes from the project and the results of the simulation will be used to identify key vulnerabilities in the current maritime systems. In terms of technical aspects, it should provide detailed information on cyber-attack scenarios and their effects, as well as a perspective on how effective and sufficient the existing maritime defence systems are, as well as a description of their strengths and weaknesses. From the human factors perspective, using the simulator, data regarding the performance of operators will be generated [52]. The integration of eye-tracking technology will allow us to better understand how operators react and make decisions when faced with these threats. The use of eye tracking can be effective in preventing errors by detecting the degradation in performance and preventing it from occurring [53]. Additionally, this can be utilised to design effective training programs.

Another outcome of the project is to identify the vulnerable points within the Turkish Straits and determine the types of cyber-attacks that are likely to occur in specific areas in the event of a cybersecurity breach. A potential direction for research would be to expand the scope of the study to include other important maritime routes beyond the Turkish Straits. By analysing these additional routes, researchers could develop a more comprehensive understanding of the global maritime cybersecurity landscape. Additionally, comparative studies could provide identification and understanding the vulnerabilities, threat profiles and impact of different geographic and operational environments on cybersecurity vulnerabilities and defence mechanisms.

The anticipated results of the project include the development of a robust AI-based risk assessment tool. The risk assessment component, supported by AI-driven tools, is expected to show the most critical points of vulnerability for cyber-attacks. The algorithms' precision and predictive power by improving their integration of a greater range of data sources can guide the development of more strong defence mechanisms and improve the overall resilience.

Based on the findings of the study, several policy recommendations will be made to national maritime authorities and shipping companies. These recommendations will include guidelines for implementing the developed cyber risk framework, strategies for enhancing human-machine interaction, and measures for improving overall cyber resilience in maritime operations. The project aims to provide actionable insights that can inform policy decisions and drive industry-wide improvements in cybersecurity.

While this study provides valuable insights, it is important to acknowledge several limitations that may have impacted the findings and their broader applicability. The study could have limitations by the small sample size due to available participant and by the as a natural consequence of short period of experiment slots [54]. Another potential challenge in this research is the accurate simulation of cyber-attack scenarios in a controlled environment [55]. Although bridge simulators are highly advanced, they may not capture all the complexities of real-world cyber threats. To mitigate this concern, the research will use historical data and expert input to develop realistic and comprehensive scenarios. Participant engagement as well as accurate data collection are also challenges during simulations. Considering the sensitive nature of the information collected, data privacy and security are also critical challenges. To ensure that all data is handled in accordance with relevant regulations, detailed data protection protocols will be implemented.

In conclusion, the expected outcomes of this research are anticipated to significantly enhance maritime cybersecurity by providing valuable insights into both the technical and human dimensions of cyber threats. The findings will inform the development of more effective training programs, defence mechanisms, and mitigation strategies, benefiting the Turkish Straits and global maritime navigation. The policy recommendations derived from this research will further support the implementation of effective cybersecurity strategies and ensuring the security and resilience of maritime operations worldwide. Also, findings can serve as a foundation for cybersecurity-focused educational approaches, such as cyber ranges, game-based learning, and online courses[16]. Through a comprehensive approach that integrates technical analysis, human factors research, and policy development, this project aims to make a substantial contribution to the field of maritime cybersecurity. Lastly, interdisciplinary studies that integrate maritime cybersecurity with economics, policy studies, and environmental science could provide a more comprehensive understanding of the broader impacts of cyber threats on global trade.

5. CONCLUSION

The objective of this research project is to combine technical analysis with human factors studies to address the complex challenge of maritime cybersecurity. Our primary objectives are to identify vulnerabilities in maritime systems, understand operator response under cyberattacks, and develop effective measures and training programs. Advanced bridge simulators will be used to simulate various cyber-attack scenarios, along with eye-tracking technology to analyse operator responses. Additionally, an AI-powered risk assessment tool will help to identify critical vulnerabilities and assist in monitoring threats and improving cybersecurity.

By the end of the project, this research is expected to provide significant benefits for the maritime industry. These measures can enhance protection and improve operational efficiency. The project is anticipated to have a significant impact on the security of the Turkish Straits, a crucial and heavily trafficked maritime route. It will enable targeted defence strategies that reduce the likelihood of cyber-attacks by identifying vulnerable points and potential attack types. Furthermore, these insights have the potential to benefit global maritime navigation by allowing for the development of global standards and best practices for maritime cybersecurity.

The research addresses key gaps in the understanding of maritime cybersecurity, contributing significantly to both the technical and human dimensions of threat management. By identifying vulnerabilities, developing targeted mitigation strategies, and providing policy recommendations, this study aims to enhance the resilience of maritime operations against cyber threats. Building on these findings, future research can further improve the security and efficiency of global maritime navigation.

6. ACKNOWLEDGEMENTS

This project is funded by the British Council through the International Science Partnerships Fund Research Collaborations programme.

7. REFERENCES

1. UNCTAD. Review of Maritime Transport 2021. 2021.

2. UNCTAD. Review of Maritime Transport 2023. 2023.

3. RAZA Z, WOXENIUS J, VURAL C A, LIND M. Digital transformation of maritime logistics: Exploring trends in the liner shipping segment. *Computers in Industry*. 2023;145.

4. UFLAZ E, SEZER S I, TUNÇEL A L, AYDIN M, AKYUZ E, ARSLAN O. Quantifying potential cyber-attack risks in maritime transportation under Dempster–Shafer theory FMECA and rule-based Bayesian network modelling. *Reliability Engineering & System Safety*. 2024;243.

5. EDMUNDS T. Navigating the Future of Maritime Cyber Security 2024 [Available from: https://www.safeseas.net/maritime-cyber-security-governance/#:~:text=It%20has%20been%20estimated%20for,over%20100%20attacks%20per%20day].

6. WINGROVE M. 14% of maritime industry hit by ransomware payments 2023 [Available from: https://www.rivieramm.com/news-content-hub/news-content-hub/cyber-security-solution-unveiled-as-maritime-incidents-rise-78685.

7. DIAKUN B. War zone GPS jamming sees more ships show up at airports 2024 [Available from: https://www.lloydslist.com/LL1148748/War-zone-GPS-jamming-sees-more-ships-show-up-at-airports].

8. AFENYO M, CAESAR L D. Maritime cybersecurity threats: Gaps and directions for future research. Ocean & Coastal Management. 2023;236.

9. Passage Planning Guide for the Turkish Straits 2024-2025 edition: *Witherby Seamanship Inter*; 2023.

10. Ministry of Foreign Affairs RoT. Note on the Turkish Straits 2022 [Available from: https://www.mfa.gov.tr/the-turkish-straits.en.mfa].

11. PAKOZ M Z, EREN F, BAS A. An analysis of the changing role of Istanbul as a megacity in the world. *Proceedings of the 55th ISOCARP World Planning Congress* 2019.

12. RUSSON M-A. The cost of the Suez Canal blockage: BBC News; 2021 [Available from: https://www.bbc.co.uk/news/business-56559073].

13. POTAMOS G, STAVROU E, STAVROU S. Enhancing Maritime Cybersecurity through Operational Technology Sensor Data Fusion: A Comprehensive Survey and Analysis. *Sensors*. 2024;24(11).

14. VU V D, LÜTZHÖFT M, IMSET M. Developing human factors design guidelines for marine electronics — the case of S-mode. *WMU Journal of Maritime Affairs*. 2023;23(1):25-48.

15. ZARZUELO I. Cybersecurity in ports and maritime industry: Reasons for raising awareness on this issue. *Transport Policy*. 2021;100:1-4.

16. SCANLAN J, HOPCRAFT R, COWBURN R, TROVÅG JM, LÜTZHÖFT M. Maritime Education for a Digital Industry. *NECESSE ErgoShip 2021–Maritime artikler*. 2022;7(1):23-33.

17. LARSEN M H, LUND M S. Cyber Risk Perception in the Maritime Domain: A Systematic Literature Review. *IEEE Access*. 2021;9:144895-905.

18. ABEYSIRIWARDHANE A. Learning and Learning-to-Learn by Doing: An Experiential Learning Approach for Integrating Human Factors into Maritime Design Education. *Maritime Technology and Research*. 2020;3(1).

19. ROSS R, PILLITTERI V, GRAUBART R, BODEAU D, MCQUAID R. Developing Cyber-Resilient Systems: A Systems Security Engineering Approach. *NIST Special Publication 800-160*. 2021;2.

20. IACS. Unified Requirement E26 Cyber Resilience of Ships. 2022.

21. IACS. UR-E27 Cyber resilience of on-board systems and equipment 2022 [Available from: https://iacs.org.uk/news/iacs-adopts-new-requirements-on-cyber-safety/].

22. MSC.428(98). Maritime Cyber Risk Management in Safety Management Systems. 2017.

23. LIEBETRAU T, BUEGER C. Advancing coordination in critical maritime infrastructure protection: Lessons from maritime piracy and cybersecurity. *International Journal of Critical Infrastructure Protection*. 2024;46.

24. DRAZOVICH L, BREW L, WETZEL S. Advancing the State of Maritime Cybersecurity Guidelines to Improve the Resilience of the Maritime Transportation System. *2021 IEEE International Conference on Cyber Security and Resilience* (CSR)2021. p. 503-9.

25. HOPCRAFT R, MARTIN K M. Effective maritime cybersecurity regulation – the case for a cyber code. *Journal of the Indian Ocean Region*. 2018;14(3):354-66.

26. YOUSAF A, AMRO A, KWA PTH, LI M, ZHOU J. Cyber risk assessment of cyber-enabled autonomous cargo vessel. *International Journal of Critical Infrastructure Protection*. 2024;46.

27. PARK C, KONTOVAS C, YANG Z, CHANG C-H. A BN driven FMEA approach to assess maritime cybersecurity risks. *Ocean & Coastal Management*. 2023;235.

28. BOLBOT V, KULKARNI K, BRUNOU P, BANDA O V, MUSHARRAF M. Developments and research directions in maritime cybersecurity: A systematic literature review and bibliometric analysis. *International Journal of Critical Infrastructure Protection*. 2022;39.

29. ERBAS M, KHALIL S M, TSIOPOULOS L. Systematic literature review of threat modeling and risk assessment in ship cybersecurity. *Ocean Engineering*. 2024;306.

30. NGANGA A, SCANLAN J, LUTZHOFT M, MALLAM S. Enabling cyber resilient shipping through maritime security operation center adoption: A human factors perspective. *Appl Ergon*. 2024;119:104312.

31. ERSTAD E, OSTNES R, LUND M S. An Operational Approach to Maritime Cyber Resilience. TransNav, the *International Journal on Marine Navigation and Safety of Sea Transportation*. 2021;15(1):27-34.

32. DOMINGUEZ-PÉRY C, ARAB K, PEREA C, TASSABEHJI R. Ship's Cyber Security: Antecedents to improve Resilience Capabilities and their Impact on Decision-making and Collaborative Performance. *The 15th Mediterranean Conference on Information Systems (MCIS) and the 6 th Middle East & North Africa Conference on digital Information Systems* (MENACIS; Madrid2023.

33. SONER O, KAYISOGLU G, BOLAT P, TAM K. Risk sensitivity analysis of AIS cyber security through maritime cyber regulatory frameworks. *Applied Ocean Research*. 2024;142.

34. HAUGLI-SANDVIK M, LUND M S, BJØRNESETH F B. Maritime decision-makers and cyber security: deck officers' perception of cyber risks towards IT and OT systems. *International Journal of Information Security*. 2024.

35. PALBAR MISAS J D, HOPCRAFT R, TAM K, JONES K. Future of maritime autonomy: cybersecurity, trust and mariner's situational awareness. *Journal of Marine Engineering & Technology*. 2024:1-12.

36. COLLIER Z A, HASSLER M L, LAMBERT J H, DIMASE D, LINKOV I. Supply chains: *Springer International Publishing*.; 2019. 447-62 p.

37. ONISHCHENKO O, SHUMILOVA K, VOLYANSKYY S, VOLYANSKAYA Y, VOLIANSKYI Y. Ensuring Cyber Resilience of Ship Information Systems. TransNav, the *International Journal on Marine Navigation and Safety of Sea Transportation*. 2022;16(1):43-50.

38. KHANDKER S, TURTIAINEN H, COSTIN A, HAMALAINEN T. Cybersecurity Attacks on Software Logic and Error Handling Within AIS Implementations: A Systematic Testing of Resilience. *IEEE Access*. 2022;10:29493-505.

39. RAJAMÄKI J, NEVMERZHITSKAYA J, VIRÁG C. Cybersecurity Education and Training in Hospitals. *Global Engineering Education Conference* (EDUCON). 2018.

40. FAN S, YANG Z. Towards objective human performance measurement for maritime safety: A new psychophysiological data-driven machine learning method. *Reliability Engineering & System Safety*. 2023;233.

41. BERGMANN B, WIEDEMANN, J., & ZEHRT, P. Description and training of competence for failure diagnosis. *Zeitschrift Fur Arbeits-Und Organisationspsychologie*. 1995;39(4):146-56.

42. WANG Z, ZHANG J, MAO Z, FAN S, WANG Z, WANG H. Emotional State Evaluation during Collision Avoidance Operations of Seafarers Using Ship Bridge Simulator and Wearable EEG. *2021 6th International Conference on Transportation Information and Safety* (ICTIS)2021. p. 415-22.

43. RONCA V, UFLAZ E, TURAN O, BANTAN H, MACKINNON S N, LOMMI A, et al. Neurophysiological Assessment of An Innovative Maritime Safety System in Terms of Ship Operators' Mental Workload, Stress, and Attention in the Full Mission Bridge Simulator. *Brain Sci*. 2023;13(9).

44. UFLAZ E. Simulator-Based Evaluation of Human Response in Emergencies: *Istanbul Technical University*; 2023.

45. MARTINEZ-MARQUEZ D P S, PANUWATWANICH K, STEWART R A, MOHAMED S. Application of Eye Tracking Technology in Aviation, Maritime, and Construction Industries: A Systematic Review. *Sensors*. 2021;21(13).

46. DUCHOWSKI A T. A breadth-first survey of eye-tracking applications. *Behavior Research Methods, Instruments, & Computers*. 2002;34 (4):455-70.

47. KITAMURA K, SETA H, OKADA K, TOBA M. Basic Study for Grasping the Human GazeMovement of a Remoting-Ship Operator. *European Journal of Maritime Research*. 2024;3(1).

48. UFLAZ E, BANTAN H F, BLACK H, ARSLAN O, AKYUZ E, TURAN O, et al. Concepts and Applications of Eye Tracking Technology in Maritime Shipboard Operation with Pilot Case Study. *23rd International Association of Maritime Universities Annual General Assembly and Conference, AGA 2023*; Helsinki2023.

49. IMO. MC 6.09 Training Course Instructors 2017

50. HART S G, STAVELAND L E. Development of NASA-TLX (Task Load Index): Results of Empirical and Theoretical Research. In: Hancock PA, Meshkati N, editors. *Advances in Psychology*. 52: *North-Holland*; 1988. p. 139-83.

51. CHITA-TEGMARK M. Attention Allocation in ASD: a Review and Meta-analysis of Eye-Tracking Studies. *Review Journal of Autism and Developmental Disorders*. 2016;3(3):209-23.

52. GARG V, VINOD G, PRASAD M, CHATTOPADHYAY J, SMITH C, KANT V. Human reliability analysis studies from simulator experiments using Bayesian inference. *Reliability Engineering & System Safety*. 2023;229.

53. PEIßL S, WICKENS C D, BARUAH R. Eye-Tracking Measures in Aviation: A Selective Literature Review. *The International Journal of Aerospace Psychology*. 2018;28(3-4):98-112.

54. REY G D. Seductive details and attention distraction – An eye tracker experiment. *Computers in Human Behavior*. 2014;32:133-44.

55. TAM K, FORSHAW K, JONES K D. Cyber-SHIP: Developing Next Generation Maritime Cyber Research Capabilities. *Conference Proceedings of ICMET Oman* 2019.

8. AUTHORS BIOGRAPHY

Yasin Burak Kurt is a researcher at the University of Strathclyde with a distinguished background in naval service. A graduate of the Naval Academy, he has a decade of experience in the Navy, where he served on various vessels, including corvettes, fast patrol ships, and oil tankers, in positions ranging from Electrical Officer to Chief Engineer. Following his naval career, he earned a master's degree in Maritime Transportation Engineering from Istanbul Technical University (ITU) and contributed to the Turkish Naval Forces' fast combat support auxiliary ship project, which provides replenishment at sea and supports intelligence gathering, search and rescue, humanitarian evacuation, and disaster relief operations. Currently, he is pursuing Ph.D. studies at the University of Strathclyde, focusing on human factors and maritime cyber security, particularly in addressing and mitigating emerging threats and vulnerabilities.

Prof. Emre Akyuz is a professor and Head of the BSc Program in Maritime Transportation Engineering at Istanbul Technical University (ITU). An ITU Maritime Faculty alumnus, he holds a PhD in Maritime Transportation Engineering from ITU and an MSc in International Transportation Management from SUNY Maritime College. With experience as an ocean-going officer and chief officer on container ships, dry-bulk cargo ships, and tankers, his research focuses on risk and reliability analysis, cybersecurity, human error prediction, and accident analysis in the maritime sector. He is recognized in the top 2% of the world's most-cited scientists by Stanford University since 2021 and received the ITU 250th Year Science Award for Best Young Academic in 2022.

Dr. Rafet Emek Kurt holds a PhD in Maritime Human Factors and is currently appointed as a Reader in Maritime Human Factors at the University of Strathclyde. Renowned as a global expert in maritime human factors, Dr. Kurt's research focuses on the design, optimization, and operation of ships, with a core emphasis on human factors, safety, and risk.

Dr. Kurt has conducted extensive research on human factors, concentrating on both the experimental and theoretical investigation of human performance and reliability in maritime and offshore operations. His areas of expertise encompass risk and safety assessments, human risk-informed design, human reliability assessment, behavioural safety, accident investigations, and human factors taxonomies. He is also deeply involved in the human-centred design of ships and offshore operations. With a significant track record in the field, Dr. Kurt continues to make impactful contributions, influencing the next generation of researchers and advancing the emerging field of maritime human factors.

Human Factors 2024
8th – 9th October 2024
Wageningen, Netherlands

ITERATING DIVERGENT AND CONVERGENT THINKING IN HUMAN-CENTRED DESIGN IN A FUTURISTIC PASSENGER SHIP BRIDGE DESIGN

Y Man, P Björndal, and **E Brorsson** ABB, Sweden
J Junkers, **E Carlson, M Lundh** and **S N MacKinnon**, Chalmers University of Technology, Sweden

SUMMARY

Human-centred design (HCD) is crucial for developing future maritime systems that meet mariners' evolving needs amidst technological advancements. However, applying HCD principles throughout the entire lifecycle remains uncommon in the maritime industry, leading to operational inefficiencies and increased risk of operator error [1]. This paper investigates how HCD practices with divergent and convergent thinking can influence the development of future maritime systems using a bridge design case. Through discover, define, develop and deliver, innovative solutions defining the next generation of maritime operational environments were developed. The insights highlight the importance of understanding user needs and user-system interactions, and the role of divergent and convergent thinking in enhancing functionality, safety, and user satisfaction of future maritime systems, contributing to the discourse on HCD application in complex system design.

1. INTRODUCTION

Human-centred design (HCD) has emerged as a crucial problem-solving approach across industries, emphasizing the early integration of end-users' needs throughout an iterative design process [2][3]. HCD frameworks have proven invaluable in enhancing user experience, efficiency, development costs and overall system performance across diverse domains [4][5][6][7], including consumer electronics, health care, manufacturing, and transportation. By placing the end-user at the centre of the design process, HCD principles ensure that humans and systems are compatible and align with the users' requirements, preferences, and operational contexts.

While HCD is a valuable broader philosophy and participatory framework, there are many barriers to its application in the maritime domain. Traditionally characterized by technology-driven innovations, the maritime industry has in recent years undergone rapid technological advancements in digitalization and automation. The fast-paced integration of these technologies within such a complex sociotechnical system often leads to insufficient understanding and consideration of end-users' needs throughout the development process. For instance, designers who often lack formal maritime education (and even vocational experience) emphasize the technical aspects of design, frequently overlooking the specific needs of onboard crew and the intricate details of a ship's working environment, paying little attention to the practical requirements of daily operations [8]. Emerged design issues may lead to operational inefficiencies, user resistance, increased risk for operator error, and ultimately, lost opportunities for successful innovative designs that respond to the evolving needs of maritime end users [9][10].

With initiatives such as the International Maritime Organization's e-Navigation and Maritime Autonomous Surface Ship (MASS) strategies [11], the evolving landscape of the shipping industry has created an increasing demand for the adoption of HCD to make sure that new technological advancements are human-centric. This shift is imperative to mitigate accidents associated with digitalisation, machine learning and automation challenges and to ensure the continued safety and efficiency of this global transport system. While the benefits of HCD are well recognized, its application throughout the entire life cycle, encompassing understanding user needs, ideation and inventive problem-solving, remains uncommon and challenging in the maritime industry [1]. This is particularly true for the development of future systems, with only limited numbers of research studies available such as [12][13][14][15][16].

To address this need, we shall consider a practical and tailored approach to HCD. This research aims to explore and communicate the entire design process, investigating how HCD practices incorporating divergent and convergent thinking can influence the development of future maritime systems. It seeks to illustrate the value of HCD through innovative solutions in a bridge design case, which, while potentially not yet technologically mature, define the next generation of maritime operational environments and provide actionable design insights for technology manufacturers and shipbuilders.

2. LITERATURE REVIEW

2.1 HUMAN-CENTRED DESIGN FRAMEWORK

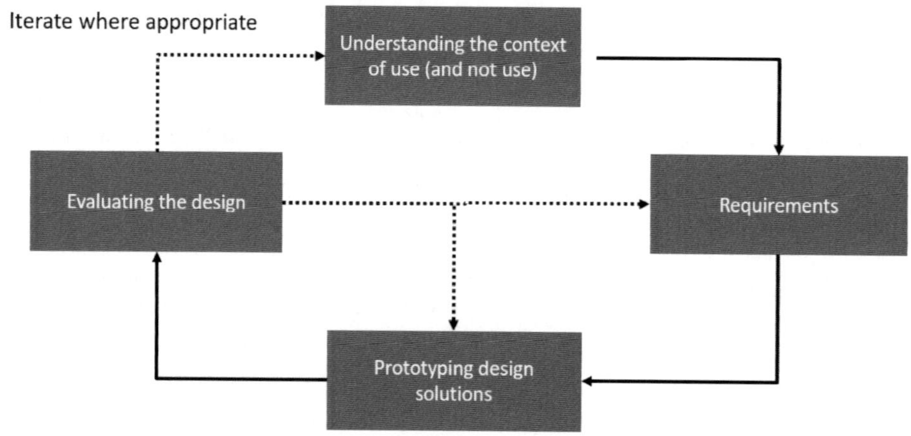

Figure 1. Human-centred design framework adapted from [17]

HCD is a systematic methodology framework to systems design and development that aims to enhance usability by focusing on system usage and applying human factors/ergonomics and usability techniques [17]. It helps identify and address users pain points, understand their needs, behaviours, experiences, and deliver customized solutions that elevate the overall user experience (UX). According to the ISO standard, HCD involves four core activities: 1) Understanding and specifying the context of use for all relevant users; 2) Identifying user needs and requirements; 3) Producing design solutions; 4) Evaluating design solutions.

By placing a strong emphasis on engaging with and empathizing users as one of the key principles, HCD facilitates finding the right problem [2]. It also helps create solutions that are not just a product or service that works but one that is intuitive and enjoyable, fitting seamlessly into the users' daily routines seamlessly. The iterative cycle [3] inherent in HCD - comprising prototyping, testing, and refining - encourages perpetual improvement, enabling solutions to adapt and evolve in response to user feedback and shifting requirements, ensuring ease of use and accessibility that contributes to a more pleasant and effective user journey.

2.2 CHALLENGES IN HCD ADOPTION IN THE MARITIME DOMAIN

HCD offers substantial benefits, particularly in improving user experience. Implementing HCD in the shipping industry requires a shift from traditional, technology-driven approaches to more user-centric methods. This shift involves overcoming several obstacles that hinder the effective integration of HCD principles into maritime design processes. Understanding these barriers is essential for devising practical application strategies that can address the unique needs and contexts of maritime operations and truly promote HCD in this domain.

2.2 (a) Conflicting User Requirements and Design Constraints:

The industry involves a wide range of stakeholders, including ship owners, operators, designers, and regulatory bodies, each with their own interests and priorities. Aligning these interests towards a common HCD goal is difficult because different groups could have conflicting requirements. Balancing these diverse requirements often necessitates a robust negotiation or mediation process in which various stakeholders are continuously engaged and concepts are iteratively designed. This dynamic could lead to prolonged design cycles and increased project costs, but it is essential for achieving a truly human-centred outcome.

2.2 (b) Lack of User Involvement or Feedback in Design Processes:

The challenge of involving users in the design process of maritime projects is multifaceted and stems from various constraints. Involving users throughout the entire life cycle of the project can be difficult. The nature of maritime operations, where users are frequently at sea for extended periods, poses logistical challenges in involving the right representatives in those design processes. The physical distance and time constraints can hinder effective communication and collaboration between designers and end-users, leading to a disconnect between design solutions and practical user needs or real pain points. Another hurdle is the difficulty in soliciting user feedback for future-oriented systems. End users, such as ship navigators and engineers, often struggle to articulate their needs in a systematic way or visions for systems that do not yet exist. Additionally, there is a notable resistance or scepticism towards advanced automation technologies among end-users. Ship navigators and engineers often exhibit a cautious attitude towards new technologies, which can hinder the collection of valuable feedback and impede user acceptance during the design phases. This scepticism is understandable given the maritime industry's longstanding reliance on proven technologies and methods, making the adoption of "untested" innovations appear risky and uncertain.

Therefore, it is crucial to develop tangible artifacts that facilitate user engagement, providing a solid foundation for gathering feedback and enhancing acceptance of new technologies. Systematic user involvement must be prioritized to ensure that feedback and needs are continuously integrated throughout the design process.

2.2 (c) Insufficient Cross-Domain Knowledge:

Implementing HCD principles requires specialized knowledge in ergonomics and human factors, areas often outside the existing skill sets of maritime professionals. Consequently, this expertise is frequently sourced from consultancies or third-party experts. However, these external consultants may lack a deep understanding of maritime practices, which can impact the effectiveness of their guidance. Additionally, while maritime regulations and codes are typically prescriptive, few explicitly mandate the integration of human factors. Developing cross-domain knowledge is crucial. It is essential for design professionals to conduct field studies and other research activities to thoroughly understand the context in which their solutions will be used.

2.2 (d) Organisational Barriers:

Shifting the focus to a more human-centric approach requires a significant cultural and organizational change, which is often met with resistance. This resistance could be rooted in the culture of the industry, which is traditionally conservative with established practices in engineering and design that prioritize technical and operational efficiencies over the needs of end-users. Such a technology-centred focus often positions the integration of human factors as a good-to-have addition, rather than a core component of system design. In the context of maritime projects, the work is usually tightly constrained by margins, and demand considerable resources - time, expertise, and financial investment. The high upfront costs of implementing HCD can deter organizations, especially those prioritizing short-term cost savings. A lack of awareness or understanding of the benefits of HCD in the management compounds this issue. Overcoming these obstacles demands education, advocacy, and tangible demonstrations of HCD's benefits to maritime operations.

2.2 (e) Technological Fragmentation:

The current state of maritime technology is characterized by a lack of integration and standardization across different systems and platforms [9][18][19]. Each system or platform may use different protocols, interfaces, and data formats, creating significant barriers to interoperability. Designing holistic HCD solutions that can operate seamlessly within this fragmented technological landscape is challenging. This fragmentation also increases the risk of human errors and inefficiencies due to the need to navigate multiple, disparate systems. Furthermore, the lack of standardization can lead to increased training requirements and maintenance complexities, as users must become proficient with a variety of tools and systems that do not necessarily communicate or function well together.

Additional factors, such as unique operational conditions and stringent regulatory framework of the maritime domain [1], further impede the adoption of HCD in shipping. To address some of forementioned challenges, particularly those that are relevant to the user involvement, requirements and knowledge gap, it is crucial to deploy practical and structured HCD strategies. This entails gaining a deep understanding of the specific challenges faced by maritime end-users, continuously engaging them in innovation and feedback throughout the project lifecycle, and supporting creative thinking and fostering iterative design processes. These efforts aim to showcase the tangible benefits of HCD, and this paper documents these processes and their outcomes through a design case study of a future ship bridge.

3. METHODOLOGY

The Double Diamond created by the UK design Council [20] is one of the practical design process guiding models for applying HCD principles for innovation. The Double Diamond consists of four phases that are aligned with HCD - Discover, Define, Develop, and Deliver, which uses a structured yet flexible approach to represent the divergent and convergent thinking that occurs during the design process (see Figure 2).

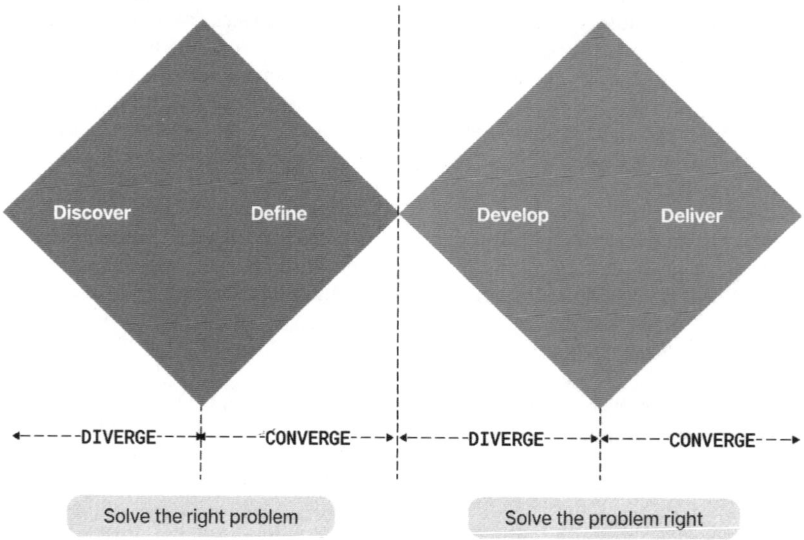

Figure 2. Double Diamond with divergent and convergent thinkings baked into two spaces - solve the right problem and solve the problem right.

First is the Discover phase, which involves broad exploratory research, using methods such as observations and interviews, to gain a comprehensive understanding of the problem space. This phase aims to uncover the full scope of the problem, identify user needs, and gather contextual insights. Then in the Define phase, the focus shifts to convergent thinking to synthesize and analyse the data collected during the Discover phase. This involves analytical method for compiling a large amount of data and drafting problem statements, to distil the broad findings into a clear and concise problem definition that serves as a solid foundation for the subsequent solution development. The Develop phase marks the transition back to divergent thinking, where we generate a wide array of potential solutions. This creative phase involves brainstorming sessions, sketching, and prototyping. The aim is to explore multiple ideas and iterate on them based on feedback and testing. The final phase, Deliver, involves convergent thinking again to refine and implement the chosen solution. This phase includes detailed testing, finalizing design elements, and preparing for the solution's handover. The goal is to ensure that the final concept addresses the defined problem.

Next, we will detail this HCD journey of Discover, Define, Develop, and Deliver phases in a future passenger ship bridge design case, manifesting how some of the HCD challenges can be overcome through the application of Double Diamond backed up by HCD principles. The case shows the whole processes from initial problem research to final design, aiming to demonstrate the tangible benefits and value that HCD can provide when appropriately adapted and applied.

4. CASE STUDY: FUTURE SHIP BRIDGE DESIGN

4.1 DISCOVER

The Discover phase of this case study focused on understanding the current state of ship bridge design and identifying user needs and pain points through extensive research and field studies. This phase involved both a selective literature review and multiple field studies on different passenger ferries to gather firsthand insights from experienced maritime professionals.

4.1 (a) Scoped Literature Review

The literature review around ship bridge design subject highlighted the outdated nature of many current ship bridges, which have seen only piecemeal technological improvements over the decades without a holistic redesign [21]. It was noted that as much as a third of all maritime accidents are attributed to inadequate design of onboard equipment [22][23]. Another study [24] described the disorganized state of ship bridge design by comparing it to a car with pedals in the back and instruments on the front passenger side, emphasizing the urgent need for a more user-centred approach.

Given the growing trends in automation and digitalization, new ship bridge designers face a complex landscape. Previous attempts at redesigning ship bridges may result in futuristic but impractical concepts that prioritize new technologies over realistic, user-centred designs [25]. To design an effective user-centred ship bridge for the future, lights have been shed on the systematic analysis, methods and criteria [15][26][27], equipment integration [28], cognitive ergonomics [29], design guidelines [30], open standards and unified approach to bridge design [14].

4.1 (b) Field Studies on Three Passenger Ferries and Bridge Simulator

The first field study was conducted on the bridge of a large Ro-Ro passenger ferry ship. This vessel, approximately 183 meters in length and capable of carrying up to 1,500 passengers and 550 cars, provided a real-world context to observe and understand the ship bridge's issues during a routine trip. Observations were made in both daylight and darkness, providing a comprehensive view of the bridge's usability under different conditions. The study included open discussions, individual interviews with bridge crews, and observations of the bridge layout and crew interactions.

A second field study was conducted on an older Ro-Ro passenger ferry. The ship is approximately 152 meters in length, capable of carrying up to 2,300 passengers and 630 cars. This daylight trip provided a comparative analysis with the aim of highlighting the similarities and differences between older and more recent designs. As with the first study, this visit included observations of the bridge layout, open discussions, and individual interviews with the crew.

The third study took place on the high-speed craft that is approximately 88 meters in length. The visit focused on its bridge, which did not have wing bridges and instead used cameras for side visibility. Although the ship was in port and not operational, the general layout of the bridge was studied, and a former captain was interviewed to provide insights into its design and functionality.

A further study was conducted at Chalmers University of Technology in Sweden, utilizing a full-scale ship bridge simulator. This environment allowed for detailed examination of navigation systems and bridge equipment without the risks associated with real-world navigation. Feedback was also gathered from an instructor who was former captain with extensive experience.

The field studies involved interviews with 11 maritime professionals, most of whom had significant experience. The group consists of four captains and three navigators, all between the ages of 40-60. The study also included younger members, two co-navigators aged 20-30 and one lookout aged 30-40. The interviews were semi-structured to explore various perceptions regarding the bridge environment. Key questions focused on understanding the current deficiencies in bridge design, the optimal layout, and the potential role of automation. Communication practices were also discussed. Looking towards the future, interviewees shared their perceptions for the bridge of the future and their thoughts on centralized layout for all controls or a more open, spread-out design.

4.2 DEFINE

The research carried out during the Discover phase unrevealed many issues in the current bridge design. One example is that the bridge workspace was often cluttered with bulky equipment, leaving little room for movement and increasing the likelihood of incidents. The off-center and stationary lookout positions, coupled with the low ceiling heights that limited vertical visibility for both lookouts and navigators, further compounded these issues.

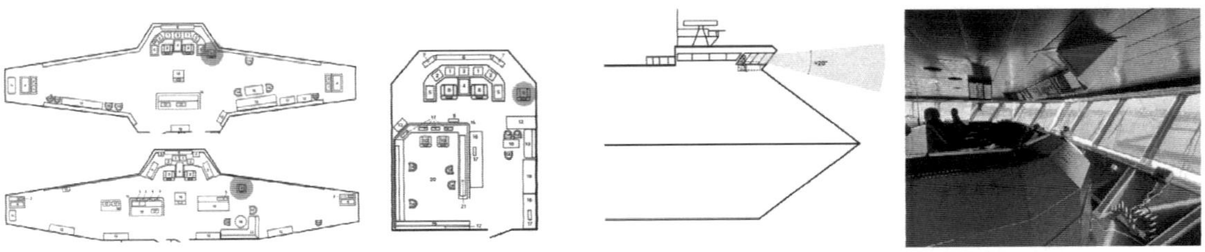

Figure 3. Off-center, stationary lookout positions (left) and low ceiling heights restrict vertical visibility (right), in a clustered bridge environment.

To systematically analyse this data from interviews and observations, the KJ method [31] was used. This method, recommended by researchers such as Kawakita Jiro, involves organizing data into clusters to identify patterns and relationships in a bottom-up fashion [32]. Findings were transferred to digital post-it notes, and then they were grouped into clusters based on similarities, and each cluster was given a name summarizing its content. The main themes that emerged from this clustering process included Night Mode, Autopilot, Feedback, Communication, Social Aspects, System Integration, General Layout, Projections, View, Circumstantial Design, Systems, Buttons, Screens, Cameras, Sit/Stand Arrangements, Bright Light, Aesthetics, and Tiredness. Each theme represented a set of related issues and insights that informed the development of design guidelines.

After clustering, the data were mapped into a list of design guidelines that focus on, e.g., bridge layout and view, incorporating user recommendations and retaining positive aspects from the existing designs. For instance, the bridge should provide an uninterrupted forward view of at least 180 degrees and good backward visibility, with windows designed to minimize reflections and refractions. The general layout should centralize navigational equipment, allow for intuitive placement following users' mental models, and provide space for personal items while ensuring easy upgradability and manual operation when needed. Information gathering should prioritize the placement of vital information in front of the navigators, ensuring accessibility without causing information overload. Physical ergonomics emphasized adjustable working positions and promoting daily physical activity to prevent tiredness.

The guidelines formulated during the Define phase provided a solid foundation for the divergent ideation activities in the Develop phase, ensuring the design process adhered to HCD principles. This approach facilitated a targeted focus on addressing critical issues identified earlier, thus enhancing the relevance of the design solutions developed.

4.3 DEVELOP

When a comprehensive understanding of the ship bridge context had been achieved and the design guidelines list completed, the Develop phase was initiated.

4.3 (a) Brainstorming and Braindrawing

The ideation consisted of structured brainstorming sessions within different focus areas that were directly connected to the findings from Define phase. The goal was to generate as many ideas as possible, emphasizing divergent thinking. For instance, initial ideas ranged from bridge shapes similar to existing designs to futuristic two-floored, 360-degree circular bridges, exploring various angles and bends of window panels. Other ideas for general bridge layout included modular controls, a central navigator, a work-table for co-navigators/lookouts for increased monitoring. Considering the visionary nature of the study, future technologies and opportunities were also explored, converting trends in different industries into bridge design ideas.

Another ideation method, braindrawing, was also used in this stage. It is basically a collaborative sketching on existing ideas to stimulate visual thinking and creativity. In this process, many ideas were generated as much as possible in the form of sketches, resulting in a wide scope of quality ideas manifested in Figure 3.

Figure 4. Various sketches for future ship bridges.

4.3 (b) Physical and Digital Prototyping

A physical prototype of the front of a ship with magnetically interchangeable bridges was constructed. This allowed for rapid prototyping of new bridge shapes and testing their functionality on this 1:50 scale model. Additionally, to assess the impact of these bridge designs on interior arrangements, 3D-printed figures and furniture were utilized to enhance the realism of the physical simulation (See Figure 4).

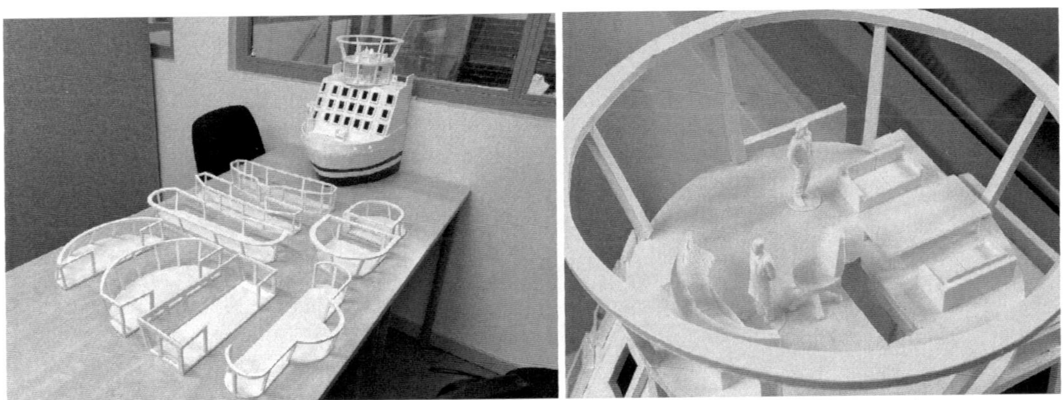

Figure 5. Physical ship model with interchangeable bridges and the 3D printed layout pieces.

Blender was also used for digital rapid prototyping, experimenting with different shapes, angles, and windows. Figma was employed for rapid prototyping of non-3D elements, such as the layout of the bridge, equipment placement and screen UI.

4.4 DELIVER

4.4 (a) Idea Evaluation, Refinement, and Concept Formulation

Around 70 initially generated ideas within 11 areas (shape, general layout, view, wing bridge, night mode, tiredness, physical ergonomics, autopilot, social aspects, communication, and information gathering) were continuously evaluated, including discussion with maritime human factors experts and end-users. This process helped narrow down the ideas to those with the most potential. Reasons for keeping or discarding ideas were carefully considered. For example, portable controls were rejected due to the risk of errors in critical situations, particularly because turning one's body with attached controls can cause disorientation, while ideas for a central navigator station were kept to optimise the conditions for the navigator.

The ideas were then further developed before being incorporated into a coherent concept. The focus here in this step is on quality over quantity. The development process involved refining ideas and making design decisions. For example, three different shapes were built digitally, evaluated, and narrowed down to one shape with outwards-leaning windows to minimize reflections and dirt buildup.

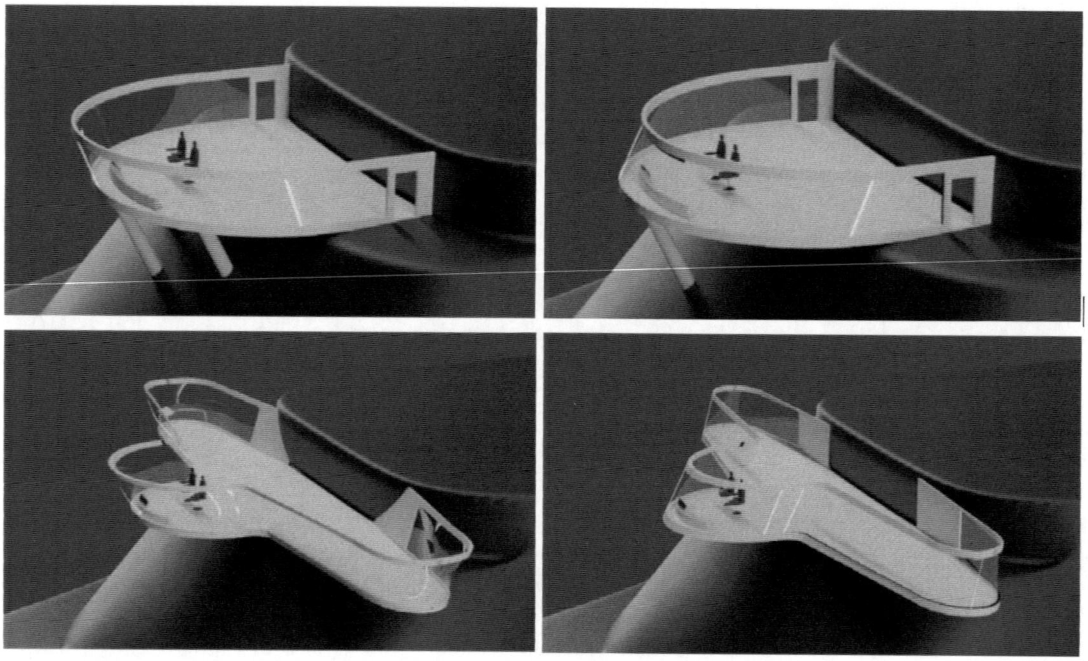

Figure 6. Evaluation of bridge shapes in Blender for making design choice. This refinement process helps with idea converging and holistic concept formulation.

Figure 7. The ideas of a central console and a wide strip of screens emerged as approaching to a coherent concept.

The reduced refined ideas were then combined into a coherent concept, iterated three times before finalization (See Figure 7-9). This iterative process involved constant evaluation and adjustments based on the experts' feedback. For example, the layout of the bridge was adjusted to remove large concave or sectioned windows and to improve backward visibility, ensuring better ergonomics and functionality.

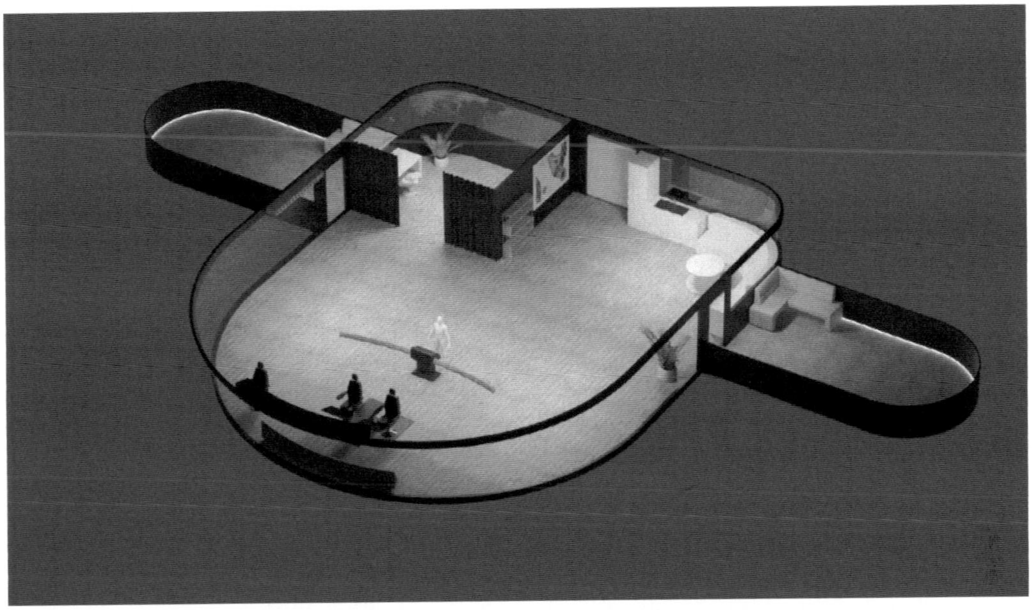

Figure 8. Initial bridge layout concept.

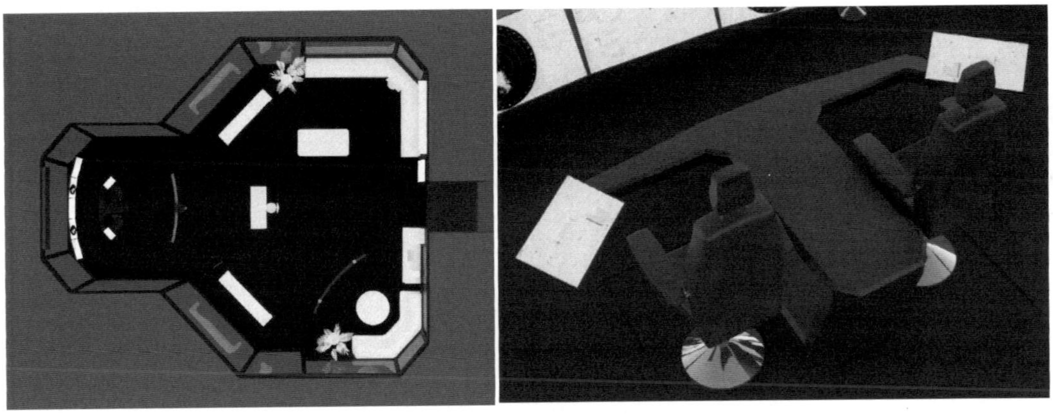

Figure 9. The second iteration of the bridge layout concept.

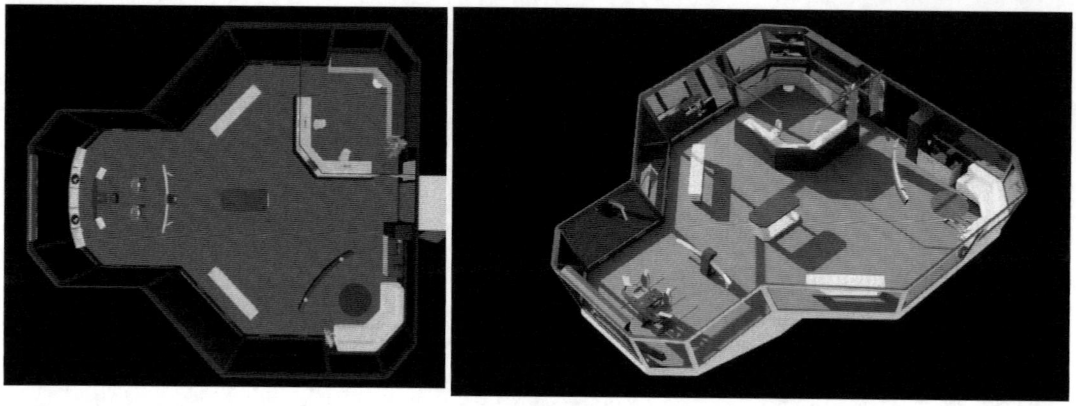

Figure 10. The third iteration of the bridge layout concept.

4.4 (b) User Evaluation and Final Concept Refinement

To assess and refine the developed concepts, continuous evaluation was carried out throughout the design process. The concepts were consistently shown and discussed with end users. Some ideas were presented orally while others in sketch format, and more finalized concepts shown as CAD renders that even allow for a precise representation of what a user would see from their position on the bridge.

Additionally, Virtual Reality (VR) technology was employed to iteratively assess the proposed concepts. Using VR headsets and hand controllers, a virtual version of the bridge was created, enabling the evaluation of component placement, size, and reachability. This method facilitated the assessment of the physical workspace from the user point of view, and thus drives iterative refinement of the design.

Upon completion of the development phase, the final concept was evaluated with users to determine whether the design guidelines had been effectively translated into the new ship bridge design. Rendered images from a CAD model were presented to 3 ex-seafarers that had significant navigational experience. During the evaluation process, users were invited to provide feedback on areas they deemed inadequate or in need of further refinement. They were also encouraged to express their opinions on colours and materials to be used on the bridge. After completing the user evaluation, the feedback collected was analysed to identify any shortcomings or areas requiring improvement in the design concept.

By iteratively integrating continuous refinement and evaluation throughout the Develop phase and Deliver phase, the design process maintained a strong focus on user needs and preferences. This ultimately enhanced the relevance and novelty of the final ship bridge design and ensure its alignment with HCD principles.

This case study can serve as a touchstone for technology manufacturers, shipbuilders, and maritime stakeholders, to understand how deep user engagement, iterative design processes, and a focus on user-system interactions can drive user-centred innovation for future maritime systems.

5. DISCUSSION

The bridge vision presented in this project sets itself apart from previous research in the domain of future bridge design by placing a greater emphasis on the physical workspace. It is crafted to facilitate both manual operations and more autonomous functions, enhancing its practicality for near-term implementation compared to other projects that may envision a more distant future. However, the project's broad scope combined with limited time constraints resulted in a visionary, yet detail-limited, ship bridge design. To further refine this study, it is crucial to undertake additional field studies to gain a deeper contextual understanding and conduct more interviews. Incorporating maritime regulations is also essential to make sure that the design adheres to compliance rules and standards, in order to strike a balance between innovation and established industry norms.

Despite the constraints shown in this project, this case study could still make a compelling argument for the wider adoption of HCD in the maritime industry, as the HCD principles were integrated at each stage of the design process. The research conducted during the Discover phase uncovered many issues in the current ship bridge design, suggesting the need for cognitive and physical ergonomics considerations. During the Define phase, the data collected from interviews and observations were systematically analysed and synthesized, leading to the development of key themes and a series of actionable design guidelines. These guidelines covered a wide range of aspects for future bridge design, such as Shape and View, General Layout, Information Gathering, etc. This phase synthesized broad findings into clear, actionable design principles. The Develop phase marked a return to divergent thinking, where extensive brainstorming and braindrawing sessions generated a multitude of innovative ideas. This creative phase was crucial for exploring various potential solutions and pushing the boundaries of traditional ship bridge design. The final Deliver phase utilized convergent thinking to refine and iterate on these ideas, incorporating continuous user feedback to ensure that the final design met user needs effectively. This iterative process drives constant refinement and validation, leading to innovative design solutions that are both functional and can truly address end user needs. For example, the design features an uninterrupted forward view of at least 180 degrees and improved backward visibility, addressing critical visibility issues identified in the Discover phase. Overall, by emphasizing understanding and integrating user needs and feedback throughout the whole life cycle of a design process and role of divergent and convergent thinking, the study showcases how HCD principles could be incorporated into complex maritime design projects.

The Double Diamond model proved to be an effective approach for structuring the design process, suitable for complex, real-world design projects. The alternation between divergent and convergent thinking was particularly beneficial. Divergent thinking creates space for exploration, while convergence brings focus, clarity, and direction. This oscillation between expansive and focused thinking is emblematic of how design thinking unfolds. It promotes the cultivation of innovative thought and creativity, which is beneficial for the conceptualization and development of future systems. This well-established methodology was adopted in our study because it is a systematic way that balances broad exploration with focused problem-solving. By employing the Double Diamond model, this study ensures that it aligns well with the principles of Human-Centred Design due to the model's inherent flexibility and concentration on the end user involvement and iteration.

Applying the Double Diamond model in design projects also introduces complexities, particularly regarding the fluidity between different phases. Often, the demarcation between phases such as Develop and Deliver is not rigid, leading to activities from earlier phases extending into later ones. Additionally, during concept development, the intention was to create at least three distinct concepts to facilitate a comprehensive evaluation of diverse ideas. However, it became apparent that the interconnected nature of the ideas would compromise the coherence of separate concepts, thus complicating user evaluation. Despite these challenges, the project successfully integrated the best ideas into a final, unified concept. This was achieved by maintaining continuity of information through meticulous documentation and employing digital tools for knowledge management and communication. Additionally, involving a wide range of stakeholders - including industrial UX experts, academic human factors specialists, end users, and ship simulator instructor - helped navigate through various HCD adoption barriers, delivering a robust and inclusive design process.

There are several promising future directions for HCD research, such as assessing the long-term benefits of HCD, examining the integration of emerging technologies such as VR, and generative artificial intelligence within HCD frameworks to understand user experience and adoption barriers, and developing assessment tools and methodologies for physical and cognitive ergonomics. These areas could further broaden HCD's impact across various industries.

6. CONCLUSIONS

The bridge design study highlights the value of HCD application in conservative sectors such as the maritime industry. By integrating HCD principles into system design, the study demonstrates the usefulness of the Double Diamond model. This model, with its iterative process and intrinsic emphasis on both divergent and convergent thinking, appears to be particularly suitable for the development of complex future systems. The design practice also highlights the continuous integration of user feedback as a key driver for the creation of practical, innovative, and user-friendly solutions. This approach is essential for the design of future technologies where user acceptance and satisfaction are pivotal to its overall success.

7. REFERENCES

1. COSTA, N. A., 'Human Centred Design for Maritime Safety: A User Perspective on the Benefits and Success Factors of User Participation in the Design of Ships and Ship Systems', *Chalmers Tekniska Hogskola (Sweden)*, 2016.

2. PREECE, J., ROGERS, Y., and SHARP, H., 'Interaction Design: Beyond Human-Computer Interaction', *John Wiley & Sons*, 2015.

3. GIACOMIN, J., 'What is Human Centred Design?', *The Design Journal*, 17(4), 606-623, 2014.

4. FARINANGO, C. D., BENAVIDES, J. S., CERÓN, J. D., LÓPEZ, D. M., and ÁLVAREZ, R. E., 'Human-Centered Design of a Personal Health Record System for Metabolic Syndrome Management Based on the ISO 9241-210:2010 Standard', *Journal of Multidisciplinary Healthcare*, 11, 21–37, 2018.

5. MALLAM, S. C., LUNDH, M., and MACKINNON, S. N., 'Integrating Human Factors & Ergonomics in Large-Scale Engineering Projects: Investigating a Practical Approach for Ship Design', *International Journal of Industrial Ergonomics*, 50, 62-72, 2015.

6. CHANG, Y., KIM, J., and JOO, J., 'An Exploratory Study on the Evolution of Design Thinking: Comparison of Apple and Samsung', *Design Management Journal*, 8(1), 22-34, 2013.

7. MÄKELÄ, M., 'Humanizing Technology', *ABB Review*, 02/2021, 2021.

8. ABEYSIRIWARDHANE, A., LÜTZHÖFT, M., PETERSEN, E. S., and ENSHAEI, H., 'Human-Centred Design Knowledge into Maritime Engineering Education; Theoretical Framework', *Australasian Journal of Engineering Education*, 21(2), 49–60, 2016.

9. MAN, Y., BRORSSON, E., and BJÖRNDAL, P., 'Human-Machine Interaction Challenges for Bridge Operations in Large Passenger Ships and Future Improvements from The Deck Officers' Perspective', *AHFE International*, 2023.

10. LÜTZHÖFT, M., and DEKKER, S., 'On Your Watch: Automation on the Bridge', *The Journal of Navigation*, 55(01), 83-96, 2002.

11. INTERNATIONAL MARITIME ORGANIZATION (IMO), Framework for the Regulatory Scoping Exercise for the Use of Maritime Autonomous Surface Ships (MASS), *IMO MSC 100/20/Add.1*, Annex 2, 2018.

12. WAHLSTRÖM, M., KARVONEN, H., KAASINEN, E., and MANNONEN, P., 'Designing User-Oriented Future Ship Bridges: An Approach for Radical Concept Design', In *Ergonomics in Design*, CRC Press, 2016, pp. 219-233.

13. COSTA, N. A., HOLDER, E., and MACKINNON, S. N., 'Implementing Human Centred Design in the Context of a Graphical User Interface Redesign for Ship Manoeuvring', *International Journal of Human-Computer Studies*, 100, 55-65, 2017.

14. NORDBY, K., MALLAM, S. C., and LÜTZHÖFT, M., 'Open User Interface Architecture for Digital Multivendor Ship Bridge Systems', *WMU Journal of Maritime Affairs*, 18, 297-318, 2019.

15. LURÅS, S., 'Systemic Design in Complex Contexts: An Enquiry Through Designing a Ship's Bridge', *The Oslo School of Architecture and Design*, 2016.

16. MALLAM, S. C., LUNDH, M., and MACKINNON, S. N., 'Integrating Participatory Practices in Ship Design and Construction', *Ergonomics in Design*, 25(2), 4-11, 2017.

17. INTERNATIONAL ORGANIZATION FOR STANDARDIZATION (ISO), 'ISO 9241-210:2019, Ergonomics of Human-System Interaction - Part 210: Human-Centred Design for Interactive Systems', Geneva: International Organization for Standardization, 2019.

18. MAN, Y., LUNDH, M., and MACKINNON, S. N., 'Managing Unruly Technologies in the Engine Control Room: From Problem Patching to an Architectural Thinking and Standardization', *WMU Journal of Maritime Affairs*, 17(4), 497-519, 2018.

19. LÜTZHÖFT, M., and NYCE, J., 'Integration Work on the Ship's Bridge', *Journal of Maritime Research*, 5(2), 59-74, 2008.

20. DESIGN COUNCIL, 'Double Diamond: A visual representation of the design and innovation process', *the Design Council*, 2004.

21. MECK, U., STROHSCHNEIDER, S., and BRÜGGEMANN, U., 'Interaction Design in Ship Building: An Investigation into the Integration of the User Perspective into Ship Bridge Design', *Journal of Maritime Research*, 6(1), 15–32, 2009.

22. GRECH, M. R., HORBERRY, T., and KOESTER, T., 'Human Factors in the Maritime Domain', In *CRC Press eBooks*, 2019.

23. DANIELSEN, B., 'The Contribution of Ship Bridge Design to Maritime Accidents', *AHFE International*, 2022.

24. BULL, M., 'Waste of Space: A Fresh Look into Bridge Design', *Safety4Sea*, 2021.

25. ROLLS-ROYCE, 'Autonomous Ships. The Next Step', *https://www.rolls-royce.com/~/media/Files/R/Rolls-Royce/documents/%20customers/marine/ship-intel/rr-ship-intel-aawa-8pg.pdf*, 2016.

26. WAHLSTRÖM, M., KARVONEN, H., KAASINEN, E., and MANNONEN, P., 'Designing for Future Professional Activity – Examples from Ship Bridge Concept Design', *AHFE International*, 2022.

27. GUO, F. B., WU, B., WAH, M., YANG, Z., BLANCO-DAVIS, E., KHALIQUE, A., and BURY, A., 'Towards An Ergonomic Interface In Ship Bridges: Identification of The Design Criteria', *AHFE International*, 2022.

28. DANIELSEN, B., LÜTZHÖFT, M., HAAVIK, T. K., JOHNSEN, S. O., and PORATHE, T., '"Seafarers should be navigating by the stars": barriers to usability in ship bridge design', *Cognition, Technology & Work*, 24(4):675-691, 2022.

29. MU, B., GUO, F. B., YANG, Z., and JENKINS, I., 'Interface and Interaction: The Symbolic Design for Bridge Conning System', *AHFE International*, 2023.

30. MALLAM, S., NORDBY, K., 'Assessment of Current Maritime Bridge Design Regulations and Guidance', *Arkitektur og Designhøgskolen i Oslo*, 2018.

31. SCUPIN, R., 'The KJ Method: A Technique for Analyzing Data Derived from Japanese Ethnology', *Human Organization*, 56(2), 233-237, 1997.

32. BLIGÅRD, L-O., 'ACD³ Utvecklingsprocessen ur ett människa-maskinperspektiv', *http://www.acd3.se/assets/files/ACD3%20-%20Utvecklingsprocessen%20ur%20ett%20manniska-maskinperspektiv%202.1.pdf*, 2015.

8. AUTHORS BIOGRAPHY

Yemao Man is a Senior User Experience Researcher at ABB's Corporate Research Centre in Sweden. His responsibilities include planning and conducting UX research to understand human-technology interactions in various industrial contexts. His previous experience includes defining future human-centred solutions for ship bridges, shore-based control, mine, pulp and paper etc. He holds a PhD in Human Factors from the Chalmers University of Technology.

Petra Björndal is a Senior User Experience Researcher at ABB's Corporate Research Centre in Sweden. Her responsibilities include applied UX research in the area of Human-computer Interaction. She holds a M.SC in Cognitive Science, and post-graduate studies in Human-computer Interaction. Previously, she has worked on human-centered solutions for industrial users in various contexts, e.g., control rooms, mines, ship bridges.

Emmanuel Brorsson holds the current position of User Experience Designer at ABB Corporate Research Centre Sweden. The Author is responsible for executing human-centred design processes focusing on experts in various domains. The Author's previous experience includes defining solutions of future maritime navigation.

Elias Carlson holds the current position of Coordinator at Aurobay through Sigma Industry West. Elias is responsible for internal tasks engine test objects and material ordering/handling. Elias has previous experience as User experience researcher as part of a master's thesis project for ABB.

Jens Junkers holds the current position of In-House Consultant at Sigma Industry West. Jens is responsible for performing and further developing tasks within the organisation. Jens has previous experience from studying Industrial Design Engineering and conducted the master's thesis within user experience research and product developing at ABB.

Monica Lundh is a part of the Maritime Human Factor Research Unit at Chalmers University of Technology, Sweden. Her main responsibility is research within work environment, human-centred design and workplace design in the light of the technological changes and development that the shipping industry has and will face. She is a former Marine Engineer, served in the Swedish Merchant Navy and holds a PhD in Shipping Technology from Chalmers, Sweden.

Scott N. MacKinnon is a Professor of Maritime Human Factors at Chalmers University of Technology. His current research is related to human-machine interactions in complex socio-technical systems. Previous to coming to Sweden, he was a Professor at Memorial University of Newfoundland where he conducted research on the impact of simulation and virtual reality upon knowledge and skill acquisition in maritime and offshore oil and gas industries. He holds a PhD in Biomedical Engineering from the University of Cape Town.

THE ROYAL INSTITUTION OF NAVAL ARCHITECTS

THE ROYAL INSTITUTION OF NAVAL ARCHITECTS

8-9 Northumberland Street, London WC2N 5DA

Tel: +44 (0)207 235 4622

events@rina.org.uk

www.rina.org.uk

Stay connected with our social platforms

The Royal Institution of Naval Architects
8-9 Northumberland Street
London, WC2N 5DA
United Kingdom

ISBN 979-8-3313-2275-5

CURRAN ASSOCIATES INC.
proceedings.com

Warship 2018: Procurement of Future Surface Vessel

London, United Kingdom
11 - 12 September 2018

The Royal Institution of
Naval Architects